In Focus – Special Book Series

Water in Industry

Editors

Yongjun Zhang, Sven Uwe Geissen and Thomas Track

IWA PUBLISHING

Published by

IWA Publishing
Unit 104-105, Export Building
1 Clove Crescent
London E14 2BA, UK
Telephone: +44 (0)20 7654 5500
Fax: +44 (0)20 7654 5555
Email: publications@iwap.co.uk
Web: www.iwapublishing.com

First published 2023
© 2023 IWA Publishing

British Library Cataloguing in Publication Data
A CIP catalogue record for this book is available from the British Library

ISBN: 9781789064148

Contents

doi: 10.2166/wst.2023.064

Editorial: 'Water in Industry' requires advanced materials and technologies to sustain the quality development

Water is a key element for many industries by acting as a feedstock, a reaction medium, an energy transfer vector, and sometimes a byproduct as well. Industry (including power generation) accounts for 19% of global water withdrawal and the portion is expected to further increase to support economic growth according to the UN World Water Development Report 2019. A certain quality of the feeding water is defined according to the roles of water in the industrial processes, which is usually ensured by properly treating the water collected from the natural resources or a municipal pipe. However, after being used in industrial units, water is usually contaminated with various organic or inorganic substances, such as raw materials, additives, and byproducts. Consequently, the industrial effluents should be further treated before being discharged into a receiving environment.

Some effluents from such industries as petroleum refining, chemical synthesizing, and paper-making often contain various types and high levels of recalcitrant organics and/or toxic heavy metals. The special properties reduce the effectiveness of conventional activated sludge which is the primary unit in most wastewater treatment processes thanks to its excellent cost-effectiveness. Accordingly, some advanced technologies have been developed and even applied at full scale, e.g. catalytic ozonation, capacitive deionization, high-performant adsorption, and membrane processes. Furthermore, industrial effluents tend to be reused in many regions which are faced with water scarcity. For example, more than 90% of industrial effluents in China are reused. Some zero-liquid-discharge plants have been constructed. Besides the challenges of conventional contaminants like nutrients and organics, the closed loop of water is faced with the problems of salts whose concentration is usually realized with the membrane and evaporators with a high input of investment and energy followed by disposal.

The IWA conference 'Water in Industry 2021' was initiated to call the attention of researchers to crack the challenges of industrial water management and to promote knowledge sharing between researchers and practitioners. The conference took place on August 23–25, 2021 and over 130 professionals attended online. Some conference papers have been collected in this book with topics covering membrane separation, adsorption, advanced oxidation, and applications. Membranes are widely integrated with the wastewater treatment process as a highly efficient separation unit which is significantly affected by fouling problems. Tong *et al.* (2021) fabricated a hydrophilic ultrafiltration membrane via grafting PVDF with *N*-isobutoxy methacrylamide and obtained a stable flux of filtrating papermaking wastewater. For an adsorption process, the regeneration of loaded adsorbents is essential. Moed & Ku (2022) compared the desorption of As(V) loaded granular activated carbon with $FeCl_3$, $CaCl_2$, and $MgCl_2$ solutions as a regeneration method. Concerning the removal of heavy metals, Guan *et al.* (2022) studied the adsorption performance of modified *Lycium barbarum* biomass under varying conditions including pH, contact time, initial concentration, and temperature. Wen *et al.* (2021) used ultraviolet and ultrasound to enhance the efficiency of ozonation for the treatment of real atrazine manufacturing wastewater in a pilot-scale reactor. Xu *et al.* (2022) developed an electro-peroxone process by fabricating a cathode with waste-tire carbon and they promoted both the yield of H_2O_2 and the current efficiency. Zhang *et al.* (2022) reported the bio-inhibition effects of biomass gasification condensate which can negatively impact the subsequent bio-digestion. The researchers investigated the efficiency of Fenton process as a pretreatment to remove the inhibition and analyzed the economic aspects. Shende *et al.* (2022) studied the removal of suspended solids, fats, oil, and grease from slaughterhouse effluents with a pretreatment process consisting of a hydra-sieve plus, a fed rotary drum filter, a skimming tank, and a dissolved air flotation. Besides the inhibitory substances and the problematic oil, the activated sludge is also faced with the problems of salts, which occurs in many types of industrial wastewater. Araújo *et al.* (2022) investigated the treatment of textile wastewater containing high salinity (up to 12.6 g·kg^{-1}) with an anaerobic–aerobic process and found that the supplementation of an easily biodegradable organic matter can be beneficial for stable performance. After the biotreatment where bulk organics and nutrients are eliminated, a further polishing treatment is usually applied to remove the recalcitrant contaminants and heavy metals. When water contains a high level of salts, corrosion and scaling might emerge as a problem for metallic equipment like containers, pumps, and pipes. Bolaji *et al.* (2022) evaluated the corrosion behavior and the scaling potentials of steel, iron, and aluminum in the oilfield-produced waters and found that the rate of corrosion followed the order: steel > iron > aluminum.

Via these papers and the speeches at the conference, we noticed a strong demand in the management of industrial waters in aspects of the treatment of toxic or refractory organics, solutions for the concentrates from water recycling, energy-efficient desalination and evaporation, onsite regeneration of loaded adsorbents, recovery of valuables, integrated and smart process control, etc. Innovative materials and technologies are required to tackle the issues and to support the sustainable development of various industrial sectors.

Guest Editors
Yongjun Zhang IWA
School of Environmental Science and Engineering, Nanjing Tech University, Nanjing, China

Sven Uwe Geissen IWA
Chair of Environmental Process Engineering, Technische Universität Berlin, Berlin, Germany

Thomas Track IWA
DECHEMA e. V., Frankfurt/Main, Germany

REFERENCES

Araújo, S., Damianovic, M., Foresti, E., Florencio, L., Kato, M. T. & Gavazza, S. 2022 Biological treatment of real textile wastewater containing sulphate, salinity, and surfactant through an anaerobic–aerobic system. *Water Science and Technology* **85** (10), 2882–2898.

Bolaji, T. A., Olumayede, E. G. & Ojo, A. M. 2022 Evaluation of corrosion and scaling potentials of oilfield waters in an offshore producing facility, Niger Delta. *Water Science and Technology* **85** (12), 3493–3509.

Guan, J., Hu, C., Zhou, J., Huang, Q. & Liu, J. 2022 Adsorption of heavy metals by *Lycium barbarum* branch-based adsorbents: raw, fungal modification, and biochar. *Water Science and Technology* **85** (7), 2145–2160.

Moed, N. M. & Ku, Y. 2022 Regeneration of As(V) loaded granular activated carbon through desorption in $FeCl_3$, $CaCl_2$ and $MgCl_2$ aqueous solutions. *Water Science and Technology* **86** (5), 1253–1268.

Shende, A. D., Dhenkula, S., Rao, N. N. & Pophali, G. R. 2022 An improved primary wastewater treatment system for a slaughterhouse industry: a full-scale experience. *Water Science and Technology* **85** (5), 1688–1700.

Tong, Y., Ding, W., Shi, L. & Li, W. 2021 Fabricating novel PVDF-g-IBMA copolymer hydrophilic ultrafiltration membrane for treating papermaking wastewater with good antifouling property. *Water Science and Technology* **84** (9), 2541–2556.

Wen, D., Chen, B. & Liu, B. 2021 An ultrasound/O_3 and UV/O_3 process for atrazine manufacturing wastewater treatment: a multiple scale experimental study. *Water Science and Technology* **85** (1), 229–243.

Xu, A., Liu, W., Chu, L., Zhang, Y., He, Y. & Zhang, Y. 2022 Enhancement of E-Peroxone process with waste-tire carbon composite cathode for tinidazole degradation. *Water Science and Technology* **85** (12), 3357–3369.

Zhang, Q., Pluschke, J. & Geißen, S.-U. 2022 Fenton oxidation as pretreatment for biomass gasification condensate: cost and biomass inhibition evaluation. *Water Science and Technology* **85** (7), 2225–2239.

doi: 10.2166/wst.2022.141

Biological treatment of real textile wastewater containing sulphate, salinity, and surfactant through an anaerobic–aerobic system

Sofia Araújo [a], Márcia Damianovic[b], Eugenio Foresti[b], Lourdinha Florencio [a], Mario Takayuki Kato[a] and Sávia Gavazza[a,*]

[a] Departamento de Engenharia Civil e Ambiental, Laboratório de Saneamento Ambiental, Universidade Federal de Pernambuco, Recife, PE, Brazil
[b] Universidade de São Paulo, Escola de Engenharia de São Carlos, Departamento de Hidráulica e Saneamento, São Carlos, SP, Brazil
*Corresponding author. E-mail: savia@ufpe.br

SA, 0000-0001-6735-4063

ABSTRACT

Real textile wastewater containing high salinity (up to 12.6 g·kg^{-1}) and surfactant (up to 5.9 mg·L^{-1} of linear alkylbenzene sulfonate – LAS) was submitted to biological treatment for colour (up to 652 mg Pt-Co·L^{-1}) and sulphate (up to 1,568.6 mg SO$_4^{-2}$·L^{-1}) removal. The influence of ethanol and molasses supplementation was firstly evaluated in anaerobic batch reactors for the removal of dyes and sulphate. Subsequently, aiming to remove aromatic amines (dye degradation by-products), an anaerobic–aerobic continuous system supplemented with molasses was applied. Supplementation had no influence on colour removal (maximum efficiencies around 70%), while it improved sulphate reduction (23% without supplementation against 87% with supplementation), and conferred robustness to the reactors, which recovered quickly after higher salinity impact. The aerobic reactor removed aromatic amines when the level of surfactants was lower than 1.0 mg LAS·L^{-1}, but the performance of the system was hindered when the concentration was increased to 5.9 mg LAS·L^{-1}. Findings suggest that the supplementation of an easily biodegradable organic matter might be a strategy to overcome wastewater fluctuation in composition.

Key words: anaerobic co-digestion, azo dyes, COD/sulphate ratio, external carbon source, interferences, LAS

HIGHLIGHTS

- The characteristics of real textile wastewater vary with the industrial process type and may contain interferences that harm biological treatment.
- Interferences are related to high levels of salinity and surfactants.
- Good treatment efficiency of colourful wastewater was obtained by applying an anaerobic–aerobic system and electron source supplementation.
- To overcome interferences we propose strategies for industrial practices.

GRAPHICAL ABSTRACT

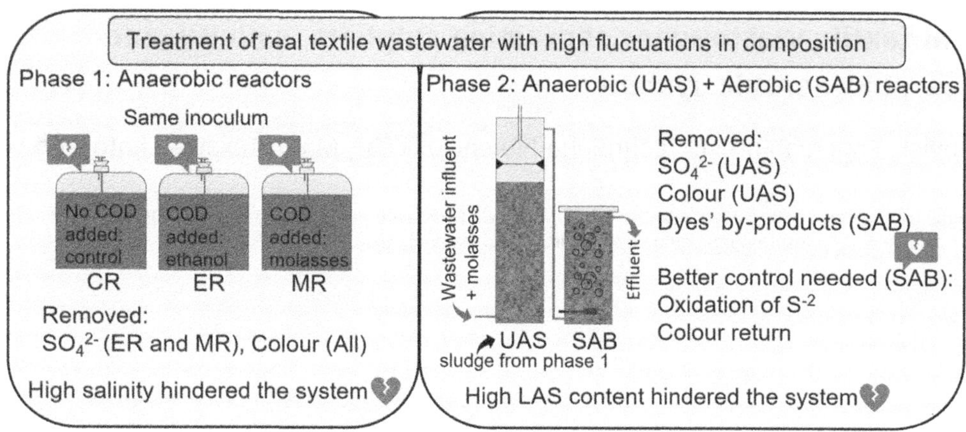

1. INTRODUCTION

Textile wastewaters are mainly characterized by high colour and salinity contents, and sometimes, high sulphate levels. Wild fluctuations are usually reported for other parameters, such as chemical oxygen demand (COD), biochemical oxygen demand (BOD) and pH (Santos *et al.* 2007).

Colour in wastewater is a result of unfixed dyes released from different industrial processes (Santos *et al.* 2007; Sudha *et al.* 2014; Köchling *et al.* 2016). As pollutants in water bodies, dyes produce undesirable aesthetic problems and can compromise water transparency (Somasiri *et al.* 2008). Additionally, they are toxic to aquatic life (Chung & Stevens 1993) and some of their degradation by-products are known to be carcinogenic and/or mutagenic (Brown & DeVito 1993).

To achieve the maximum fixation of the dyes to the fibre, high concentrations of salts, such as sodium nitrate, sodium sulphate and sodium chloride, are used during the industrial process (Carliell *et al.* 1998). High salinity content in the wastewater can cause negative effects on microbes, compromising biological treatments (Woolard & Irvine 1995).

Due to the chemical characteristics of azo dyes, a common group of dyes used in the textile industry (Santos *et al.* 2007), anaerobic followed by aerobic technologies have been applied for the biological treatment of this kind of wastewater (van der Zee & Villaverde 2005; Pandey *et al.* 2007; Silva *et al.* 2012; Cirik *et al.* 2013; Menezes *et al.* 2019; Oliveira *et al.* 2020; Carvalho *et al.* 2022). Under anaerobic conditions, reduced azo dyes can generate several types of aromatic amines, which may represent a toxic and recalcitrant potential at this stage of the treatment (Carliell *et al.* 1998; Santos *et al.* 2007; Singh & Arora 2011). Aromatic amines, conversely, are well known to be degraded through aerobic processes (Pinheiro *et al.* 2004).

Regarding sulphate content, when in high concentration and under anaerobic conditions, sulphate-reducing bacteria (SRB) may compete with the consortium of methanogenic archaea (MA) for electron donors (Damianovic & Foresti 2007; Callado *et al.* 2015). In the context of textile wastewater, dyes and sulphate are both electron acceptors and may compete for the electrons available from organic matter (Santos *et al.* 2007).

It is known that a COD/sulphate ratio (COD/SO$_4^{2-}$) of 0.67 is required to achieve complete organic matter removal through sulphate reduction (Lens *et al.* 1998). At higher ratios, other electron acceptors must be available to achieve complete organic matter removal, for instance dyes. For complete sulphate removal, electron donors and carbon sources must be added to sulphate-rich wastewaters which are deficient in COD (Liamleam & Annachhatre 2007).

Amaral *et al.* (2014) reported a prevalence of sulphate over dye reduction for the anaerobic treatment of a real textile wastewater. These authors reported that the supplementation with external electron donors may be a strategy to achieve success in both sulphate and colour removal. Hao *et al.* (2014) stated that different electron donors can provide different benefits and drawbacks to the processes, and the choice of the best electron source would depend on the reaction requirements. However, the evaluation of the behaviour of different electron sources for this competition applied to the anaerobic treatment of a real textile wastewater is still scarce.

Liamleam & Annachhatre (2007) highlighted in their review that ethanol was an attractive electron donor for sulphate reduction and that the complete oxidation of ethanol into CO_2 by SRB alone had already been reported. Together with ethanol, molasses was also chosen as an electron donor for the present study due to its low cost and great availability in Brazil, from sugarcane industry. Lactate, a fermentation by-product of molasses, is known to be a common substrate for SRB (Muyzer & Stams 2008). Two major groups of SRB can reduce sulphate into sulphide by the incomplete oxidation of lactate into acetate and CO_2, and by the complete conversion of acetate into CO_2 or HOC_3^- (Chen *et al.* 2008).

Regarding the presence of surfactants, this class of compounds is found in softening chemicals used by the industry, and they are also part of chemicals used to improve the binding of the dyes to fibres (Yaseen & Scholz 2019). Interestingly, as far as we know, surfactants are not reported as interference in studies that apply biological treatments of textile wastewater. However, the presence of surfactants and their by-products used in softening processes can impair biological treatment, causing undesirable effects on aquatic life and microbial ecosystems (Motteran *et al.* 2017).

The aim of the present study was to contribute to a better practical understating of an anaerobic–aerobic system applied to the treatment of a real textile wastewater, containing high concentration of salt, sulphate, surfactants, and colour. Considering that most studies in this field were conducted using synthetic effluents, we intend to broaden the comprehension on how those chemical compounds interfere in the biological treatment of a real harsh wastewater. In our study, we also propose strategies to overcome the fluctuations in wastewater content in order to improve the biological treatment performance.

2. MATERIALS AND METHODS

In this study, two phases of a bench-scale experiment were performed to treat a real textile wastewater with high sulphate and salinity content and a mixture of different dyes. In the first phase, under anaerobic conditions, sequential batch reactors were operated to evaluate the influence of two different external electron donors and carbon sources (ethanol and molasses) on sulphate and colour removal.

During the second phase, the complete treatment was applied: the anaerobic stage, consisting of an up-flow anaerobic structured-bed reactor with molasses as external electron donor and carbon source, followed by the aerated stage (submerged biofilter reactor). In addition, the effect of linear alkylbenzene sulfonate (LAS) was observed.

2.1. Real textile wastewater

Real textile wastewater was the substrate of this study. It was provided by a textile laundry industry, which performed the process of jeans degumming, bleaching, dyeing, and softening using many chemical compounds. The textile wastewater was stored in an equalization tank and treated by a physicochemical process in the industry. The substrate for the reactors used in this study was collected approximately every 15 days from the equalization tank of the industry.

2.2. Reactor and operating procedure

2.2.1. Phase 1 – monitoring of the batch reactors

Three glass batch reactors (CR, ER and MR) with 4 L or dm^3 of working volume and 2 dm^3 of exchange volume were essayed (Figure 1(a)). Reactor CR was the control reactor, in which no external electron source was added. For the others, a COD of 1,200 mg $O_2 \cdot L^{-1}$ was supplemented with ethanol (ER) or molasses (MR). The cycle time was 24 hours (15 minutes of manual feeding, 23 hours of reaction, 30 minutes of settling, and 15 minutes of manual discharging). The reactors were inoculated with 2 g volatile suspended solids (VSS)$\cdot L^{-1}$ (mixed liquor content) of anaerobic sludge from a pilot-scale upflow anaerobic sludge blanket (UASB) reactor, which was used to treat real textile wastewater (Amaral *et al.* 2014), resulting in a COD/biomass ratio of 0.6 of g $O_2 \cdot L^{-1}$/g VSS. No pre-treatment was applied to the sludge before using it as inoculum. It was just homogenized prior use. The reactors were kept at 30 °C (303 K), in a shaker (Innova, WEG) at 1.25 Hz of orbital agitation.

2.2.2. Phase 2 – monitoring of the continuous reactors

Two continuous-flow reactors were used as an up-flow anaerobic structured-bed reactor (UAS) followed by a submerged aerated biofilter (SAB) (Figure 1(b)). Both reactors (0.55 and 0.35 m in height) were filled with expanded clay balls (2 cm in diameter) to avoid biomass losses, resulting in a working volume of 3 and 2 L or dm^3 for UAS and SAB, respectively.

Real textile wastewater supplemented with 1,000 mg $O_2 \cdot L^{-1}$ of COD using molasses was pumped into the UAS reactor at a flow rate of 2 mL$\cdot$min^{-1} (hydraulic retention time – HRT – of 1 day). The UAS reactor was inoculated with a mixture of anaerobic sludge from the three previous batch reactors (CR, ER and MR), resulting in a mixed liquor content of 5.5 g

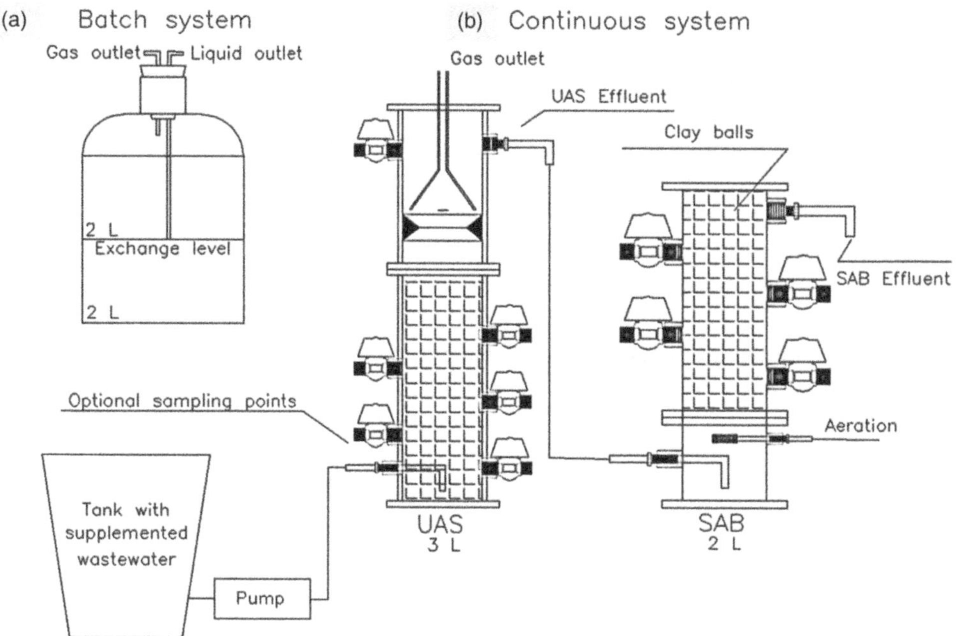

Figure 1 | Diagram of the experimental system and reactors used in phase 1 (a) and in phase 2 (b).

VSS·L^{-1} and a COD/biomass ratio of around 0.2 g O$_2$·L^{-1}/g VSS. The aerobic reactor was not inoculated. The dissolved oxygen (DO) in the UAS effluent was below the detection limit during the whole experimental period, while in the effluent from SAB, the DO was increased to 0.16 ± 0.1 mg O$_2$·L^{-1} by using an aquarium pump installed at the entry of the aerobic reactor. The anaerobic–aerobic system was maintained at 30 °C (303 K).

2.3. Monitoring parameters

The analytical procedures were conducted following the Standard Methods for the Examination of Water and Wastewater (APHA 2005). Eight lots of wastewater were collected from the industry and characterized in terms of chemical oxygen demand (COD) (SM 5220D), sulphate (SM 4500SO$_4^{2-}$E), alkalinity (SM 2320B), BOD (BOD$_{5,20}$) (SM 5210BODB), total nitrogen TNK (SM 4500NorgB), ammonia (SM 4500N-NH$_3$C), phosphorous (SM 4500P D), and colour (SM 2120 C). Salinity and pH were also measured (Hach – HQ40d). For colour determination, the light absorption peak in the VIS spectrum (Hitachi U-2910, Hitachi, Ltd, Tokyo, Japan) was at 667 nm (phase 1) and 573 nm (phase 2); thus, the absorbance at these wavelengths was chosen as an indirect measurement of the sum of different amounts and kinds of dyes in the real wastewater used in phase 1 and phase 2, respectively. Linear alkylbenzene sulfonate was measured as described by Souza *et al.* (2016).

The performance of the reactors was evaluated for 72 days (phase 1) and 75 days (phase 2), by measuring COD, sulphate, sulphide (4500S^{2-}F), colour (absorbance), alkalinity, pH, oxidation–reduction potential (ORP) (*Hach – HQ40d*), and dissolved oxygen (Hach – HQ40d). The values measured for sulphide were discounted from the COD values according to the theoretical relation described by Chan & Farahbakhsh (2015), in which 1,000 mg·L^{-1} of sulphide (as S) confers 1,996 mg·L^{-1} of oxygen demand (as O$_2$). Aromatic amines were qualitatively evaluated during phase 2 by scanning in the range from 200 to 350 nm (Pinheiro *et al.* 2004) twice a week.

At the end of phase 1, the temporal profiles were performed in ER and MR (phase 1), for the COD/SO$_4^{2-}$ ratios of 0.67 and 1.50. The first profile occurred 1 week after the day 72 of phase 1; during the 7 preceding days, both reactors were daily fed with the target COD supplementation and the lot 5 of real wastewater. The same was done for the second profile, which occurred 2 weeks after the end of phase 1. The days before each profile were not monitored. The first analyses were conducted right after feeding the reactors, while the following were conducted every 2 hours, during 24 hours of a typical batch cycle. The COD/SO$_4^{2-}$ ratios were based on the COD added (ethanol or molasses) and the sulphate content in the real wastewater.

For both phases, the efficiencies of reactors were calculated considering the influent wastewater and the effluent of each reactor CR, ER, and MR (phase 1), and UAS and SAB (phase 2). This means that efficiencies removal observed in SAB also represent the global efficiency of the entire system. In this study we present mass concentrations as $mg \cdot L^{-1}$ or $g \cdot L^{-1}$, which correspond to $10^{-3} \, kg \cdot m^{-3}$ or $1 \, kg \cdot m^{-3}$, respectively.

3. RESULTS AND DISCUSSION

3.1. Characteristics of the textile wastewater

Table 1 shows the characterization of the eight lots of wastewater used throughout the experimental period (phases 1 and 2). The textile wastewater content was considerably variable, due to the different processes performed by the industry, in which different chemicals are used. The concentrations ranged from around 500 to 1,600 mg $SO_4^{-2} \cdot L^{-1}$, for sulphate; from 4 to 13 $g \cdot kg^{-1}$, for salinity; from 250 to 1,000 mg $O_2 \cdot L^{-1}$, for COD; and from 168 to 652 mg $Pt \cdot Co \cdot L^{-1}$ for colour, resulting in 0.083 and 0.463 of minimum and maximum absorbance (both at 667 nm, during phase 1), respectively. The pH ranged from 5.6 to 7.6; when lower than 6.0, it was adjusted to 7.0 using sodium bicarbonate before feeding the reactors. During the second phase, the concentration of LAS was increased from less than 1 to up to 5.9 $mg \cdot L^{-1}$. An important fact to consider in the textile wastewater characterized here is the low content of nitrogen and phosphorous. Based on the recommended ratio of 250:5:1 for COD:N:P for anaerobic treatment (Metcalf & Eddy 2003); the eight lots of wastewater used during this study presented sufficient concentrations of nitrogen, but insufficient concentrations of phosphorous in lots 1, 2, 3, 4, 5, and 8. Supplementation of the later macronutrient would be needed to provide more favourable conditions for the microbial community to growth; however, in this study, we assessed only the supplementation of organic matter, to avoid increasing costs to the industry.

Table 1 | Characterization of the eight lots of wastewater used in the study, collected from the industry

	Lots of wastewater collected from the industry							
	1 (phase 1)	2 (phase 1)	3 (phase 1)	4 (phase 1)	5 (phase 1)	6 (phase 2)	7 (phase 2)	8 (phase 2)
COD (mg $O_2 \cdot L^{-1}$)	497.5 ± 167.5	472.2 ± 167.9	694.1 ± 214.7	991.1 ± 176.8	245.0 ± 61.4	299.4 ± 138.1	416.7 ± 25.3	838.2 ± 54.8
COD/SO_4^{2-}	0.56 ± 0.2	0.83 ± 0.3	0.63 ± 0.2	0.71 ± 0.1	0.16 ± 0.1	0.34 ± 0.18	0.79 ± 0.05	1.55 ± 0.21
Sulphate (mg $SO_4^{-2} \cdot L^{-1}$)	896.1 ± 27.6	573.9 ± 97.4	1,101.0 ± 29.1	1,413.3 ± 374.8	1,568.6 ± 50.8	863.5 ± 84.8	529.4 ± 39.5	547.6 ± 74.7
Colour (mg $Pt \cdot Co \cdot L^{-1}$)	168	652	399	462	570	382	345	502
Bicarbonate alkalinity (mg $CaCO_3 \cdot L^{-1}$)	55.9 ± 2.7	196.8 ± 17.5	0	0	301.8 ± 28.5	237.4 ± 44.3	80.1 ± 22.9	111.9 ± 92.1
Total alkalinity (mg $CaCO_3 \cdot L^{-1}$)	84.2 ± 7.4	307.6 ± 37.6	109.5	117.3	427.3 ± 75.5	347.0 ± 58.9	145.7 ± 40.6	254.4 ± 172.2
Salinity ($g \cdot kg^{-1}$)	7.2 ± 0.2	6.2 ± 0.2	7.5 ± 0.1	12.6 ± 0.2	6.2 ± 0.1	4.7 ± 0.1	3.7 ± 0.1	5.8 ± 0.2
pH	6.8	7.0	5.6	5.7	6.6	7.2	7.0	7.6
$BOD_{5,20}$ (mg $O_2 \cdot L^{-1}$)	60	140	120	580	160	70	100	310
Nitrogen TKN (mg $N\text{-}TKN \cdot L^{-1}$)	14.6	14.3	15.5	20.5	21.0	14.7	15.5	22.2
Ammonia (mg $N\text{-}NH_3 \cdot L^{-1}$)	5.8	6.6	5.8	5.3	6.8	3.4	2.7	7.1
Phosphorus (mg $P\text{-}PO_4^{2-} \cdot L^{-1}$)	0.1	0.8	0.6	0.1	0.3	1.7	3.0	1.9
COD:N:P	250:7.3:0.1	250:7.6:0.4	250:5.6:0.2	250:5.2: < 0.1	250:21.4:0.3	250:12.3:1.4	250:9.3:1.8	250:6.6:0.6
LAS ($mg \cdot L^{-1}$)	<1	<1	<1	<1	<1	<1	5.9	2.9

3.2. Phase 1 – monitoring of the batch reactors

Buffer was added during the start-up period of MR, due to the quick fermentation of molasses, by applying the $NaHCO_3$:COD ratio of 1:1. Effluent pH of either ER or CR was always higher than 6.0, thus external buffering was not needed (Table 1 of Supplementary Material). In general, the performance behaviour of ER and MR was similar; considering supplementation, influent COD/SO_4^{2-} ratio varied from 1.0 to 3.7 during the monitoring period, with no lag phase for both COD and sulphate removal.

Considering the entire experimental time, average efficiencies for COD removal were $25 \pm 24\%$, $43 \pm 15\%$, and $43 \pm 16\%$ for CR, ER and MR, while the corresponding values for sulphate removal efficiency were $8 \pm 7\%$, $47 \pm 15\%$, and $44 \pm 14\%$. The highest efficiencies were found after the 25th day of operation, with COD removal of 73 and 68% in ER and MR, respectively (Figure 2(a)), and corresponding values of sulphate removal of 74 and 70% (Figure 2(b)). In CR, an 18-day lag phase period was found for COD removal, achieving the maximum removal efficiency of 79% on the 42nd day of operation (Figure 2(a)). For sulphate, no significant removal was found throughout the operational period in CR, being barely removed from the 29th day (Figure 2(b)).

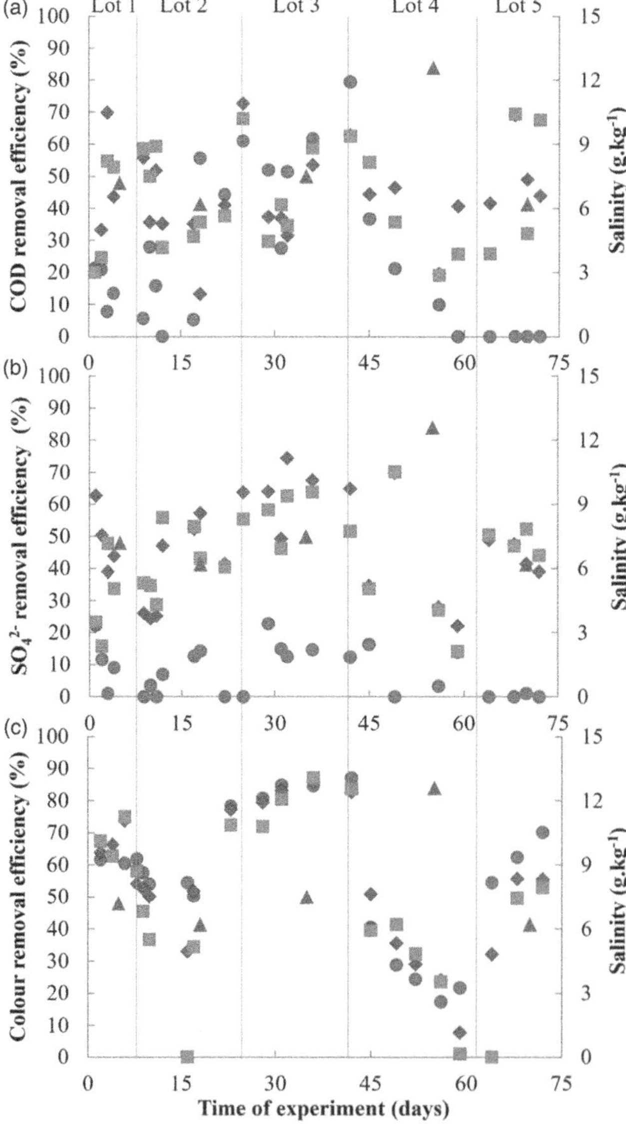

Figure 2 | Efficiencies of COD (a), sulphate (b) and colour (c) removal in CR (●), ER (♦) and MR (■) effluents, and the salinity (▲) of each lot of wastewater used in phase 1.

As it was not possible to measure and analyze the outlet gas from the reactors, some calculations contributed to better understand the COD removed by sulphidogenesis and methanogenesis. The amounts of COD used for sulphidogenesis could be theoretically inferred by using the COD/sulphate ratio of 0.67 for complete organic matter removal through sulphate reduction. The COD amounts used for colour removal could be calculated as well, by using the relation between a tetra-azo dye Direct Black 22 (DB22) and COD proposed by Amorim *et al.* (2013), explained later. Methanogenesis was considered to be responsible for the remaining amounts of COD removed.

It is interesting to note that, during the period of best performance of ER and MR, when lot 3 was used to feed the reactors, the COD removed was almost equally directed to sulphidogenesis and to methanogenesis (around 50% for each). Out of the average COD removal efficiency of 49% in both reactors, sulphidogenesis and methanogenesis represented 50 and 50% in ER and 48 and 52% in MR, correspondingly. That represents a good equilibrium among these microbial populations reflecting an adaptation since the start of the reactors' operation. The situation was changed with lot 4 of textile wastewater when the salinity increased from 7.5 to 12.6, dropping the COD removal efficiency to 38 and 34% in ER and MR, with corresponding percentage derived to sulphidogenesis to 36 and 39%. The values for each of the five lots can be found in Table 2 of Supplementary Material.

Sulphide measured was, on average, 30.5 ± 32.7, 95.2 ± 30.8 and 78.0 ± 24.6 mg S^{2-}.L in CR, ER and MR effluent, respectively. The values of sulphide for each of the five lots and the theoretical amount of H_2S released can be found in Table 3 of Supplementary Material.

Based on the literature, the SRB have the ability to use various electron donors for growth, but polymeric organic compounds are not direct substrates for them; instead, low molecular weight substrates, such as lactate, formate, and ethanol (or hydrogen), are the compounds usually oxidized by SRB, as is the case of the common *Desulfovibrio* and *Desulfotomaculum* species (Liamleam & Annachhatre 2007; Muyzer & Stams 2008).

Molasses, which comprises a high amount of sucrose (Annachhatre & Suktrakoolvait 2001), is first fermented by microorganisms such as lactobacilli into by-products that act as electron donors and a carbon source for SRB, such as lactate or pyruvate (Maree *et al.* 1986; Maree *et al.* 1987). *Desulfovibrio* and *Desulfomicrobium* species ferment pyruvate to form acetate, carbon dioxide and hydrogen as products. In efficient hydrogen-consuming systems (by methanogens), these microbes can also oxidize lactate and ethanol into acetate (Muyzer & Stams 2008). Conversely, ethanol alone, or as a fermentation product of the anaerobic degradation of carbohydrates, can be completely oxidized by SRB (Liamleam & Annachhatre 2007).

As degumming is commonly conducted by the textile industry, starch may be available as a carbon source, but it is not a direct substrate for SRB (Muyzer & Stams 2008). In CR, although the COD/SO_4^{2-} ratio (Table 1) was close to the stoichiometric value of 0.67, sulphate removal was very low, as previously presented. This indicates that the COD from the textile wastewater may not be easily degradable or easily available for SRB, which was confirmed by the low $BOD_{5,20}$ values compared with those of COD (Table 1).

Regarding colour removal, apparently, the microorganisms were less dependent on the availability of an easily degradable carbon source. Removal efficiencies for CR, ER and MR were, on average, $57 \pm 20\%$, $55 \pm 21\%$, and $47 \pm 27\%$, respectively. The three reactors achieved 86% of colour removal around day 35 of the experiment (Figure 2(c)). In CR, colour removal around 61% was found on the 8th day, achieving 87% of maximum efficiency around day 40. The colour removal efficiency in CR deserves attention.

Amorim *et al.* (2013) correlated colour, measured by mg Pt-Co·L^{-1}, and the concentration of DB22, widely used in the textile industries. According to the authors, 400 mg Pt-Co·L^{-1} corresponded to 65 mg·L^{-1} of DB22. In the present work, if we consider that the measured colour in the influent of lot 5 (570 mg Pt-Co·L^{-1}) is derived from the dye DB22 (4 azo bonds – the worst scenario), it would correspond to 92.6 mg DB22·L^{-1} of dye concentration. The complete removal of 1 mol of DB22 (1,083.97 g) requires 16 mol of electrons (four electrons for each azo bond), thus resulting in 1.37×10^{-3} mol·L^{-1} of electrons required for the complete colour removal of 92.6 mg DB22·L^{-1} present in the last lot of phase 1. Considering molecular oxygen reduction to water (4 mol of electron for 32 g of O_2), the COD required for the complete dye removal would be 10.94 mg O_2·L^{-1}, which only represents 2% of the COD measured in lot 5 (490 mg O_2·L^{-1} – Table 1); therefore, the low COD value required for dye removal in CR justifies the good colour removal even with little or no apparent COD removal.

The use of molasses as the electron donor for colour removal is not commonly reported in the literature, while ethanol has been frequently used. Menezes *et al.* (2019), treating synthetic textile wastewater and using ethanol as the carbon source, found COD removal efficiency of 76% and DB22 removal efficiency of 81% during the steady state of their study.

Finally, regarding alkalinity and ORP behaviour, COD and sulphate removal was always linked to both parameters. In the supplemented reactors, the production of bicarbonate alkalinity (the average values during lot 3 were 1,007.0 and 1,214.4 mg CaCO$_3$·L^{-1} for ER and MR, respectively) and ORP close to −400 mV were observed on almost all days of analysis. In CR, the best sulphate removal efficiency (23%) was detected around day 30 (Figure 2(b)), when the bicarbonate alkalinity increased from 489.5 to 1,183.7 mg CaCO$_3$·L^{-1} and when ORP achieved −389.5 mV (Table 2).

Although ORP of around −200 mV has been cited as the standard for sulphate reduction through H$_2$S formation (Santos *et al.* 2007), in this study, ORP close to −400 mV was found when the treatment achieved the best removal efficiencies of COD (Figure 2(a)), sulphate (Figure 2(b)) and colour (Figure 2(c)). Similar results were found by Amaral *et al.* (2014), who obtained a methanogenic environment in a UASB reactor treating real textile wastewater (average ORP of −357 mV).

3.2.1. Salinity influence

The salinity of the 4th lot of wastewater was 12.6 g·kg^{-1}, which is about one-third of the seawater salinity (35 g·kg^{-1}) (Ibrahim & Attia 2015). The increase in salt concentration was a consequence of the high amount of sodium chloride and sodium metabisulphite, used in dyeing and bleaching processes, respectively, by the industry, which were predominantly performed among several other processes when the wastewater was generated.

The 4th lot was firstly used on the 44th day of operation and immediately caused a decrease in the removal efficiencies of the three reactors. Considering the last day of monitoring of the 3rd and the 4th lots, COD removal dropped from 79% to 0% in CR, from 63% to 40% in ER, and from 63% to 27% in MR, while for colour the corresponding values went from 87% to 22%, from 83% to 3% and from 84% to 0%; and for sulphate, the values stayed below 10% and dropped from 65% to 12% and from 52% to 14% in CR, ER and MR, respectively. During the days for lot 4, average values of the COD removed used for sulphidogenesis and methanogenesis indicated that the first was harmed the worse by the increase in salinity. Corresponding values of 36 and 64% were found in ER, and 39 and 61% in MR, out of the 38% (ER) and 34% (MR) of COD removal efficiency for the period (Table 2 of Supplementary Material). Additionally, the reactors' self-buffering capacity diminished expressively when fed with lot 4 (Table 2).

In the 5th lot, the salinity of around 6 g·kg^{-1} relieved the osmotic pressure in the reactors. ER and MR recovered immediately; sulphate and colour removal were increased to around 50%, the production of bicarbonate alkalinity was detected again. CR did not recover as ER and MR (Figure 2, Table 2). The percentage of the COD removed used in ER increased to 54% for sulphidogenesis and decreased to 45% for methanogenesis. While in MR, the sulphidogenesis was responsible for 69% out of around 50% of COD removed against 31% for methanogenesis (Table 2 of Supplementary Material).

Regarding dye removal, the inoculum used in this study (Amaral *et al.* 2014) was well adapted to treat textile wastewater with high colour content (mix of different dyes) and relatively high salinity (up to 4.4 g·kg^{-1}). The same biomass was also used

Table 2 | Wastewater lot number, salinity (g·kg^{-1}), ORP (mV), pH range, and bicarbonate alkalinity (BA) (mg CaCO$_3$·L^{-1}) in the influent and in CR, ER and MR effluents during phase 1

| | | Influent | | CR effluent | | ER effluent | | MR effluent | |
| | | pH = 6.5–7.4 | | 6.2–7.4 | | 6.4–7.6 | | 5.9–7.1 | |
Lot – salinity	Day	ORP	BA	ORP	BA	ORP	BA	ORP	BA
1 – 7.2	3	11.7	57.8 (321.7[a])	82.3	178.7	−361.4	309.9	−305.6	282.1
2 – 6.2	10	4.2	214.3	82.5	144.9	−325.5	385.7	−309.7	211.5
	18	8.0	179.3	−164.5	222.7	−387.7	495.9	−345.0	289.7
3 – 7.5	30	−24.1	489.5	−389.5	1,183.7	−403.5	1,354.3	−391.1	1,458.0
	43	−12.4	864.0	−379.4	1,237.7	−408.2	2,013.1	−402.1	2,324.2
4 – 12.6	46	−335.5	598.3	−372.2	760.3	−409.9	1,773.4	−406.5	1,913.8
	59	35.6	2,246.4	−319.2	1,737.8	−389.7	2,241.2	−402.5	2,116.4
5 – 6.2	65	−120.6	319.7	−265.3	175.0	−399.7	922.3	−398.2	868.3
	69	−22.4	324.0	−11.1	118.8	−397.8	887.8	−399.2	788.4
	73	22.7	261.6	12.5	172.2	−402.1	817.5	−398.4	797.9

[a]Bicarbonate alkalinity in MR influent after the addition of the buffer.

by Köchling *et al.* (2016); in their study, they found an adapted azo dye-degrading microbial community with the ability to treat a real textile wastewater under the hash condition of salinity (up to $10\,g\cdot kg^{-1}$), which justifies the good recovery of colour removal efficiency in CR. Nonetheless, the inoculum was not able to recover the sulphate removal behaviour, as it had been previously exposed to a lower sulphate content (below $500\,mg\,SO_4^{-2}\cdot L^{-1}$) (Amaral *et al.* 2014) than that found in lot 5 ($1{,}568.6 \pm 50.8\,mg\,SO_4^{-2}\cdot L^{-1}$), reinforcing the dependence of SRB on favourable conditions, which was promoted in ER and MR by the addition of ethanol and molasses, respectively.

3.2.2. Temporal profiles and kinetic study

Although the reactors did not reach the steady state during our experimental period, the temporal profile performed with ER and MR gave us a better view of what was happening into both reactors during one cycle (24 h) and under two different COD/SO_4^{2-} ratios (0.67 and 1.5).

In the first profile (Figure 3) the COD/SO_4^{2-} ratio of 0.67 was the stoichiometric ratio for sulphate removal by the consumption of the external carbon source added (ethanol or molasses), without considering the COD content from the textile wastewater itself (average of $245\,mg\,O_2\cdot L^{-1}$ for lot 5).

The COD removal efficiencies were 63 and 69%; while sulphate removal efficiencies were 23 and 32% for ER and MR, respectively. This indicated that the sulphate reduction pathway was not dominant for organic matter removal; only 23 and 29% of the COD removed were used for sulphate reduction, in the respective reactors. Methanogens were players too, considering the ORP values found during the monitoring period (Table 2) and the low COD needed for colour removal, as presented before.

Nevertheless, in the second profile, when more ethanol or molasses were added at the ratio of 1.50, SRB found better developing conditions (98% removal for ER and 81% for MR), which may be due to a temporal accumulation of intermediate substrates. This is a consequence of higher energy conservation of fermentative microbes compared with that of the

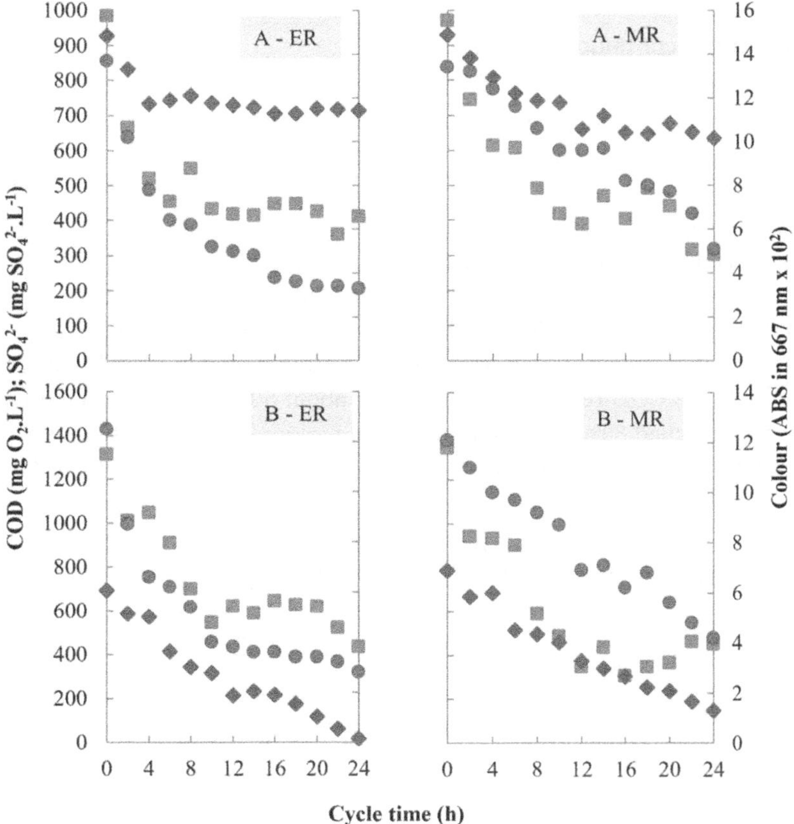

Figure 3 | Temporal profiles performed in ER and MR for COD (●), sulphate (■) and colour (values of absorbance are multiplied by10^2) (◆) during phase 1 with COD/SO_4^{2-} equal to 0.67 (a) and 1.5 (b).

methanogens (Thauer *et al.* 1977; Muyzer & Stams 2008). Out of the 67% of COD removed in both reactors (around 900 mg $O_2 \cdot L^{-1}$), 51% in ER and 48% in MR were used for sulphate reduction. These values are similar to those found during the period when the best performance of phase 1 (lot 3) was observed, in which sulphidogenesis and methanogenesis were in equilibrium.

Regarding the kinetics analysis of sulphate, the degradation observed only in the first hours of the cycle made it unreasonable to calculate and compare its removal rate due to an unreliable adjustment in both reactors when the COD/SO_4^{2-} ratio was 0.67. When the electron availability was increased ($COD/SO_4^{2-} = 1.50$), it was possible to verify that different electron sources had no influence on the removal rate (Table 3).

Conversely, colour removal was stable in both profiles with removal around 76% for ER and 63% for MR (Figure 3). A zero-order kinetic adjustment in MR suggested that the removal rates were the same under the two COD/SO_4^{2-} ratio conditions (Table 3).

Some authors have observed a first-order kinetic adjustment for the decolourization by mixed anaerobic sludge as a process dependent on the residual azo dye concentration or the electron transference, and not on the production of the reducing equivalent (van der Zee *et al.* 2001; Menezes *et al.* 2019; Oliveira *et al.* 2020). However, none of the studies has considered several electron acceptors being reduced simultaneously, as reported here, with a high amount of sulphate and a mix of various dyes.

Contrary to previous reports, our zero-order kinetic adjustments, now considering both sulphate and colour reduction rate, indicated that the production of the reducing equivalents was the limiting factor. This was reported in the case of pure cultures (van der Zee *et al.* 2001), with reactions occurring at a constant rate without dependence on electron concentration (Levenspiel 1987).

Regarding colour removal rates in the ethanol reactor, it must be considered that H^+ production in ER occurred probably faster than in MR, due to the molasses fermentation process. Thus, rapidly available H^+ in ER may have permitted colour degradation at a higher rate already in the first hours of the cycle, resulting again in an unreliable adjustment of zero-order kinetic in ER.

3.3. Phase 2 – monitoring of the continuous reactors

Table 4 brings ORP, bicarbonate alkalinity, and the average values of pH and dissolved oxygen through the phase 2 in both UAS and SAB reactors. During the continuous experiment, ORP variation and the average values of DO showed the environmental changes from an oxidant wastewater influent in the UAS reactor (DO = 0.43 ± 0.4 mg $O_2 \cdot L^{-1}$ and ORP = -5.9 ± 30.8 mV), to an anaerobic condition in its effluent (DO below the detection limit and ORP = -372.5 ± 72.9 mV); and further, returning to a less reductant condition in the SAB effluent (DO = 0.16 ± 0.1 mg $O_2 \cdot L^{-1}$, ORP = -115.4 ± 79.0 mV). The effluent from the UAS reactor showed pH near 7 during the entire experimental period (Table 1 of Supplementary Material).

Considering the addition of molasses as an external carbon source, the wastewater used during the second phase presented a COD/SO_4^{2-} ratio of 1.5 ± 0.2; 2.7 ± 0.3; and 3.5 ± 0.4 for lots 6, 7 and 8, respectively.

Figure 4 shows the COD, sulphate, and colour removal efficiency values found in the effluent of UAS and SAB. Considering the entire experimental period, The COD removal efficiency was about 50% in UAS during the first 20 days (ORP close to -400 mV, Table 4), whereas it reached 80% in SAB, in the same period; however, this can be attributed to the adsorption process, as the aerated biofilter was not inoculated. The average removal efficiencies for UAS were $57 \pm 16\%$ for COD,

Table 3 | Values of kinetic constant, kinetic adjustment and kinetic order applied to the removal of colour (in MR, when $COD/SO_4^{2-} = 0.67$ and $COD/SO_4^{2-} = 1.50$) and sulphate (in ER and MR when $COD/SO_4^{2-} = 1.50$)

	$COD/SO_4^{2-} = 0.67$		$COD/SO_4^{2-} = 1.50$	
	ER	**MR**	**ER**	**MR**
Sulphate	Not calculated due to low removal efficiencies		$k_0 = 30.66$ mg$\cdot$L$^{-1}\cdot$h^{-1}; $R^2 = 0.9460$ Zero order	$k_0 = 29.17$ mg$\cdot$L$^{-1}\cdot$h^{-1}; $R^2 = 0.9648$ Zero order
Colour	Not adjustable to 'zero order' kinetics	$k_0 = 0.0031$ mg$\cdot$L$^{-1}\cdot$h^{-1}; $R^2 = 0.9706$ Zero order	Not adjustable to 'zero order' kinetics	$k_0 = 0.0034$ mg$\cdot$L$^{-1}\cdot$h^{-1}; $R^2 = 0.9685$ Zero order

Table 4 | Wastewater lot number, salinity (g·kg^{-1}), LAS concentration (mg·L^{-1}), ORP (mV), bicarbonate alkalinity (BA) (mg CaCO$_3$·L^{-1}), pH range and DO average values (mg O$_2$·L^{-1}) in the influent, in UAS and SAB effluents during phase 2

		Influent		UAS		SAB	
		DO 0.43 ± 0.4		BDL		0.16 ± 0.1	
		pH 6.2–7.8		6.4–8.4		7.6–8.7	
Lot – Salinity – LAS	Days	ORP	BA	ORP	BA	ORP	BA
6 – 4.7 – <1	1	−19.7	236.2	−393.5	292.0	47.0	474.0
	6	−12.6	208.2	−398.7	560.0	−22.4	444.0
	13	−10.1	242.0	−419.1	596.0	−85.2	426.0
	22	−24.9	228	−409.1	944.0	−125.0	990.0
	28	−55.0	161.2	−423.2	1,198.0	−154.6	959.2
	36	−58.0	312.4	−426.2	1,313.4	−158.8	1,333.3
	42	−10.3	274.0	−432.6	1,106.0	−12.2	922.0
7 – 3.7 – 5.9	49	15.2	56.1	−404.4	496.0	−125.8	546.0
	57	12.3	73.3	−323.6	118.4	−226.7	310.2
	59	12.3	111.9	−162.3	65.8	−146.2	210.6
8 – 5.8 – 2.9	67	52.8	203.9	−329.7	172.3	−209.9	320.8
	71	27.1	19.8	−347.8	275.2	−165.5	455.4

BDL, below the detection limit.

48 ± 19% for sulphate, and 26 ± 21% for colour. While considering the entire system (measured in the effluent of SAB), efficiencies were 74 ± 14% for COD, 37 ± 12% for sulphate, and 26 ± 21% for colour removal, respectively.

Highest COD removal efficiencies were achieved around day 28, reaching 94% in the final effluent of the two reactors, with ORP close to −430 mV and −200 mV in UAS and SAB effluents, respectively (Figure 4(a), Table 4). In SAB, the biomass started to self-develop around the 12th day, indicated by the decrease in ORP values (from −14 to −100 mV) because of the DO consumption by bacteria (Kato *et al.* 1997).

Regarding sulphate removal, an increase in efficiency was observed after the first month of experiment (Figure 4(b)). At around day 45, sulphate removal efficiency in UAS was 65%, reaching a maximum of 82% during the first days of lot 7. In UAS, the amount of the COD removed used for sulphidogenesis and methanogenesis achieved 43 and 57%, respectively, during the use of lot 7 (Table 4 of Supplementary Material). Sulphide measured in the anaerobic reactor stayed around 106.3 ± 43.5 mg S^{2-}.L. The values of sulphide for each of the three lots can be found in Table 5 of Supplementary Material.

In SAB, due to aeration, part of the sulphide generated in the anaerobic reactor was reoxidized to sulphate and the global efficiency of the system decreased to 50%, considering sulphate removal. Between days 30 and 45, the sulphide measured in the UAS effluent was 150 ± 46 mg S^{2-}.L^{-1}, while in the SAB effluent, it decreased to 8 ± 17 mg S^{2-}.L^{-1}. Carvalho *et al.* (2022), treating a synthetic textile wastewater containing salinity (2.84 g·kg^{-1}), DB22 (45 mg·L^{-1}), starch (1,000 mg O$_2$·L^{-1}), and sulphate (209 and 1,270 mg SO$_4^{2-}$.L^{-1}), attributed sulphide oxidation in their aerated continuous stage to the futile cycle of sulphur. In the case of the anaerobic continuous stage, the authors discussed that for the chemical reduction of azo bonds, the biogenic sulphide served as an electron donor. According to Van der Zee *et al.* (2001), in an anaerobic environment rich in sulphide, the chemical reduction of an azo dye will be part of the decolourization process; thus, this has likely occurred in the present experiment, but it was not further investigated.

In contrast with observations in phase 1, colour removal efficiency reached 65% only after 13 days of experiment in UAS. In SAB, adsorption may have been a mechanism for colour removal during the first days (Figure 4(c)). This was also observed by Ferraz *et al.* (2011), treating real textile wastewater in a similar anaerobic–aerobic system.

Until the day 15, colour removal in UAS, indicated by the lower peak at 573 nm, was associated with the formation and accumulation of aromatic amines, observed in the range from 200 to 350 nm (Figure 5(a)). At the same day, the aromatic amines formed in the UAS reactor were not removed in the SAB reactor, which is another evidence that the aerobic biomass was not established yet.

Between days 28 and 45, the anaerobic reactor presented the maximum colour removal efficiency of 81% (Figure 4(c)). During the days of best global performance, the colour removal that occurred in UAS was followed by the removal of aromatic amines in SAB (Figure 5(b)). Nevertheless, despite biomass activity in SAB, part of the removal of aromatic amines

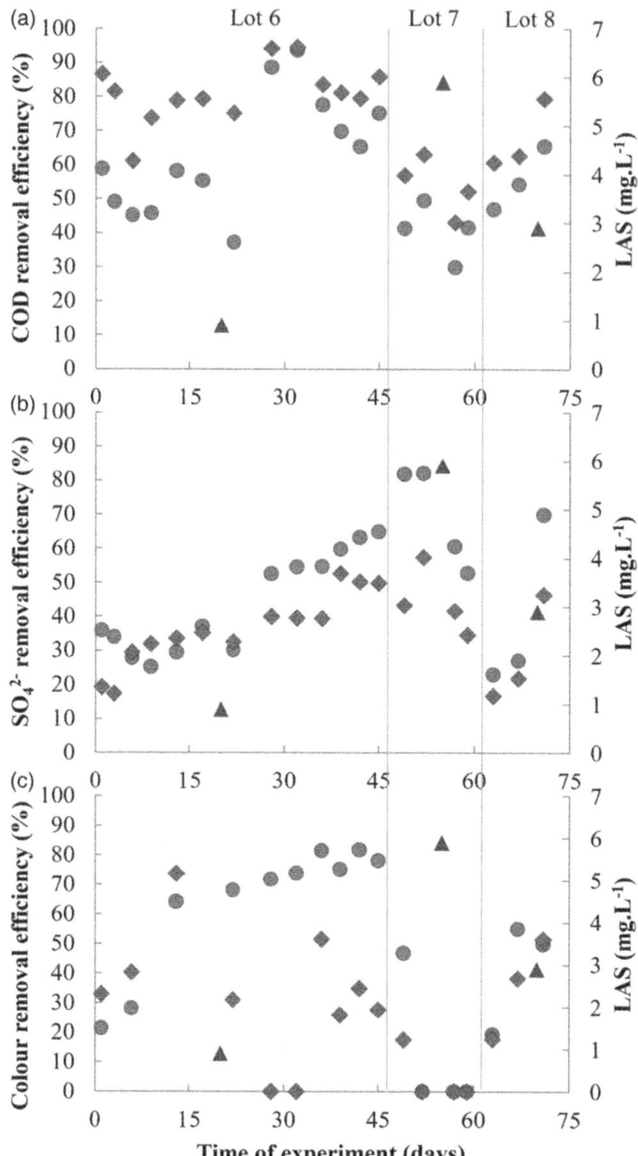

Figure 4 | Removal efficiencies of COD (a), sulphate (b) and colour (c) in UAS (●) and in SAB (◆) effluents, and LAS concentration (▲) (mg·L^{-1}) in each lot of wastewater used in phase 2.

may be a consequence of their abiotic autoxidation, which led to a decrease in colour removal efficiency. The compounds formed from autoxidation reactions resulted in an increase in the absorbance values, with the absorbance peak close to the influent characteristic wavelength (573 nm) (Figure 5(b)). Sometimes the peak was even greater than that of the wastewater's, bringing the colour removal efficiency equal to zero (Figure 4(c), days 28 and 32). This was observed in spectrophotometric scans of many days of monitoring (data not shown).

In some studies using a synthetic substrate, the same decrease in colour removal efficiency was observed in the aerobic stage of a similar system (anaerobic–aerobic stages) for the treatment of dyes (van der Zee *et al.* 2001; Jonstrup *et al.* 2011; Menezes *et al.* 2019; Carvalho *et al.* 2022). This can be explained by the tendency of some azo dyes present for autoxidation when exposed to aerobic conditions, forming coloured polymeric structures that are recalcitrant for biological treatment (Jonstrup *et al.* 2011).

Considering a real textile wastewater, some previous studies that applied a continuous anaerobic–aerobic system have also found comparable results in terms of removal efficiencies. Ferraz *et al.* (2011) achieved 77 and 86% of global removal

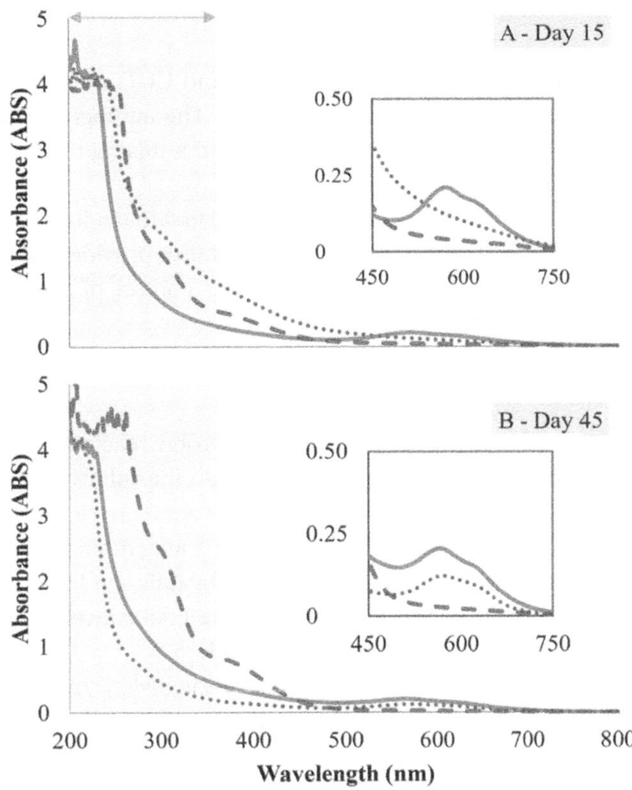

Figure 5 | Spectrophotometric scans of the real wastewater influent (——) and effluents from UAS (– –) and SAB (······) on days 15 (a) and 45 (b) of operation during phase 2.

efficiencies of COD and colour, respectively, when HRT was 24 h. The anaerobic reactor represented 59 and 64% of removal of the same parameters. Contrarily to our findings, these authors did not mention any decrease in colour removal efficiency when the anaerobic effluent was exposed to oxygen. Regarding sulphate removal, the study cited reached efficiencies between 14 and 63%, but the concentration of the compound was much lower (between 14.5 and 52.0 mg $SO_4^{2-}.L^{-1}$) than the concentrations found in the present study.

Amaral *et al.* (2014) obtained an average COD removal efficiency between 56 and 71% for the entire system. Despite the low average removal efficiencies of sulphate (varying from 41 to 54%) in the anaerobic reactor, the authors reported that there were sufficient electrons for the complete removal of the compound (varying from 269 to 464 mg $SO_4^{2-}.L^{-1}$), indicated by the COD/SO_4^{2-} ratio between 2.4 and 3.0. In addition, as we found in our study, they observed sulphide oxidation under aerobic conditions.

In the two studies cited above, the presence of salts was mentioned as a possible interference for biological treatment efficiency. The values of salinity varied from around 1.6 to 19 $g \cdot kg^{-1}$ at first (Ferraz *et al.* 2011), and they were up to 4.4 $g \cdot kg^{-1}$ in the second work (Amaral *et al.* 2014). In both studies, HRT was varied but no electron supplementation was performed; and high sulphate content impaired colour removal under anaerobic conditions. This occurrence differed from the findings of our study, maybe due to a prioritized use of electrons for sulphate reduction in their experiments.

Using a similar up-flow anaerobic structured-bed reactor similar to ours and HRT of 24 h, Florêncio *et al.* (2021) treated a synthetic wastewater containing azo dye and ethanol as the energy source. With a COD/SO_4^{-2} ratio equal to 5 and a sulphate concentration of 48 mg $SO_4^{-2} \cdot L^{-1}$, average removal efficiencies were 78, 77 and 68% for COD, sulphate and colour, respectively. The authors highlighted the good performances for colour and organic matter removal in the presence of sulphate ions and stated that sulphidogenesis did not outcompete methanogenesis nor significantly impaired reductive decolourization.

In our study, we did not observe any evidence of competition between dye-reducing and SRB even with a high content of sulphate (>500 SO_4^{-2} $mg \cdot L^{-1}$). By contrast, dye reduction occurred simultaneously with sulphate reduction. This may be explained by the supplementation of an easily biodegradable organic matter (molasses) and the low electron demand for dye reduction, as discussed previously.

Using only a UASB reactor, Somasiri *et al.* (2008) achieved colour and COD removal efficiency of around 97 and 90%, respectively, during a 24-h HRT and with no carbon supplementation. The authors called for attention to the necessity of a degradable carbon and energy source for microbes, and they suggested a mixing of textile with food factory or municipal wastewater for low organic load before feeding the anaerobic reactor.

Based on these previous works and in our results, it is possible to infer first that a low sulphate and salt contents, aligned to an available organic matter which is easily biodegradable, is fundamental to provide a more favourable condition for a biological anaerobic–aerobic treatment system; and second, that a better control of the oxidation reactions under aerobic conditions (second stage of this treatment strategy) is crucial.

3.3.1. Influence of LAS

On day 49, excessive foaming was observed in the upper part of the anaerobic reactor. LAS analysis revealed a concentration of 5.9 mg LAS·L^{-1} in the wastewater (lot 7), around six times greater than the values of the previous lots (always lower than 1 mg.L^{-1}) (Table 1).

Changes in the wastewater constituents may cause a disturbance in the anaerobic reactor and, consequently, in the entire system. Souza *et al.* (2016) analyzed methanogenesis inhibition caused by different levels of LAS present in domestic wastewater and found a reduction of 30% in methanogenesis activity when the LAS concentration was increased from 0 to 10 mg LAS·L^{-1}, and 50% when the surfactant concentration was 30 mg LAS·L^{-1}.

In our study, the anaerobic biomass had never faced a LAS concentration higher than 1 mg LAS·L^{-1}; thus, a small increase in the surfactant content caused the observed disturbance. During the use of lot 7, between days 49 and 59, considering the anaerobic reactor, ORP reached -162.3 mV, the alkalinity was consumed (Table 4), pH dropped to the lowest value of 6.4, and COD, sulphate, and colour removal efficiencies decreased to 30, 53 and 0%, respectively (Figure 4).

Lot 7 was then replaced by lot 8, which presented a LAS concentration again higher than the usual (2.9 mg LAS·L^{-1}). Although the production of bicarbonate alkalinity did not exceed 300 mg CaCO$_3$·L^{-1} (Table 4), the system indicated a recovery when considering COD and colour removal efficiencies already on the first day of analysis. An increase of sulphate removal efficiency was observed only on the last day of lot 8 (Figure 4), which is in accordance with the 15% of COD removed used for sulphidogenesis against 85% for methanogenesis (Table 4 of Supplementary Material). This indicates that the SRB activity was again worse impaired and that they would probably need more time to re-establish themselves in the UAS behaviour. The values of ORP were close to -350 mV (Table 4), similar to those observed by Amaral *et al.* (2014) (-357 mV on average) and still within the range for azo dye reduction, from -100 to -500 mV (Santos Cervantes & van Lier 2007).

3.4. General aspects of continuous and batch reactors' performance

In the two phases of this study, the best removal efficiencies of COD, colour and sulphate under the anaerobic condition were related to values of ORP around -400 mV and the production of bicarbonate alkalinity above the value of 300 mg CaCO$_3$·L^{-1}. This concentration of bicarbonate alkalinity is the recommended value for self-sufficiency of anaerobic treatment of domestic wastewater (Metcalf & Eddy 2003), which is comparable with the textile wastewater treated in our study in terms of low organic load.

Yet considering anaerobic reactors, when a continuous system was applied (phase 2), the treatment achieved COD and sulphate maximum removal efficiencies were greater than the corresponding values found for the batch system (phase 1, MR). In the UAS reactor, the maximum COD and sulphate removal efficiencies were 94 and 82%, compared with 68 and 70% in MR, respectively. In addition to the higher salinity presented by the lots used during phase 1, fresh wastewater with harsh compositions added into the system every day, with an exchange volume of 50% may have caused a daily impact to the microbial community.

Another important fact is the sulphide accumulation in the batch reactor during its cycle time, which may have inhibited bacterial activity at some level. Visser (1995) suggested anaerobic filter reactors can tolerate higher concentration of sulphide than reactors with suspended sludge. Chen *et al.* (2008) bring different aspects related to sulphide toxicity, including the total and unionized forms of the compound, their concentration in the environment, pH range, among others. The authors emphasize that different levels of sulphide can cause inhibition to several trophic groups, in different anaerobic degradation steps.

Additionally, they suggest that in reactors with fixed biomass, methane-producing microorganisms (MPM) can better tolerate sulphide due to an adaptation to free H_2S.

In our study, under the continuous regime (phase 2), the constant gas (unionized form, H_2S) and effluent outflow (ionized form as HS^-) against the exchangeable volume of 50% in batch reactors (phase 1) may have softened the toxicity generated by the sulphur compounds, resulting in a more favourable environment for the microbial community.

With regards to the COD used by SRB and MPM, Visser (1995) found that SRB and MPM can fairly compete under pH values between 6.9 and 7.7, by performing the same range growth rate and being equally inhibited by sulphide. Indeed, in the present work, during the days of best performance of both systems, the organic matter removed through sulphidogenesis and methanogenesis was around 50% each (in ER and MR, lot 3, and in UAS during the last days of lot 6 and first days of lot 7) (Tables 1, 2 and 4 of Supplementary Material).

Nevertheless, the sulphide measured in the anaerobic reactors (lower than 150 mg S^{2-}.L during the entire experimental period, in both phases 1 and 2) was not in a inhibition range in comparison to the salinity and the LAS contents. The microbial community was hindered by these two wastewater constituents, mainly the SRB group, which was able to remove less amount of COD when both batch and continuous systems suffered the impact of the respective interfering compounds.

Finally, apart from all these circumstances, the short period of feeding the reactors with each lot of wastewater, with fluctuating composition, prevented the microbes to establish themselves under such conditions, and thus the reactors to reach the steady-state. In this study, the goal was to evaluate the biological treatment of a real wastewater with minimum action to strengthen the microbial community, for instance the supplementation of phosphorous, which represent an increase in the cost of treatment.

4. CONCLUSIONS

The supplementation of an external carbon source improved the performance of the anaerobic reactors used to treat a real textile wastewater rich in sulphate, salinity, and colour by providing a more friendly characteristic to the influent wastewater. The addition of ethanol and molasses had no influence on dye transformations, but it improved sulphate reduction, being necessary when both dye and sulphate removal were targeted along with organic matter removal. Furthermore, the addition of molasses or ethanol gave robustness to the reactors, which quickly recovered from an adverse increase in salinity (from 7.5 to 12.6 $g{\cdot}kg^{-1}$).

Aromatic amines and sulphide were formed as a result of dye and sulphate reduction in the anaerobic reactor, respectively. Nevertheless, the oxidation of these compounds needs to be better controlled in the aerobic stage, since it is directly related to the overall efficiency of the treatment.

Finally, aiming to achieve an efficient and stable biological treatment, we propose that the textile wastewater must be monitored and managed before being sent to the reactors. In order to treat a less hostile influent in the reactors (lower sulphate values, availability of an easily degradable organic material and fewer interferences, such as salt and surfactants), the wastewater from different predominant industrial processes should be mixed in a way to minimize the high concentration of the interferences. Furthermore, the supplementation with an external carbon and electron source, easily biodegradable, must be considered as a strategy to achieve high treatment efficiency.

ACKNOWLEDGEMENTS

For the financial support, we thank CNPq (National Council for Scientific and Technological Development. Brazil) (grant to Sávia Gavazza, process number 304862/2018-5 and grant to Sofia Araújo, process number 165130/2017-2), and FACEPE (Science and Technology Foundation of the State of Pernambuco. Brazil) (grant to Sávia Gavazza. process number APQ 0456-3.07/20), and CAPES (Coordination for the Improvement of Higher Education Personnel) for the Graduate Program support (Proap) and Institutional Internationalization Program (PrInt) (grant to Sofia Araújo, process number 88887.363316/2019-00). We also thank the Mamute Textile Industry for wastewater provision during the study.

DATA AVAILABILITY STATEMENT

All relevant data are included in the paper or its Supplementary Information.

REFERENCES

Amaral, F. M., Kato, M. T., Florêncio, L. & Gavazza, S. 2014 Color, organic matter and sulfate removal from textile effluents by anaerobic and aerobic processes. *Bioresource Technology* **163**, 364–369. doi:10.1016/j.biortech.2014.04.026.

Amorim, S. M., Kato, M. T., Florêncio, L. & Gavazza, S. 2013 Influence of redox mediators and electron donors on the anaerobic removal of color and chemical oxygen demand from textile effluent. *Clean – Soil, Air, Water* **41** (9), 928–933. doi:10.1002/clen.201200070.

Annachhatre, A. P. & Suktrakoolvait, S. 2001 Biological sulfate reduction using molasses as a carbon source. *Water Environment Research* **73** (1), 118–126. doi:10.2175/106143001(138778.

APHA 2005 *Standard Methods for the Examination of Water and Wastewater*, Encyclopedia of Forensic Sciences: Second Edition, 21st edn. American Public Health Association, Washington. doi:10.1016/B978-0-12-382165-2.00237-3.

Brown, M. A. & DeVito, S. C. 1993 Predicting azo dye toxicity. *Critical Reviews in Environmental Science and Technology* **23** (3), 249–324. doi:10.1080/10643389309388453.

Callado, N., Damianovic, M. H. R. Z. & Foresti, E. 2015 Resilience of methanogenesis in an anaerobic reactor subjected to increasing sulfate and sodium concentrations. *Journal of Water Process Engineering* **7**, 203–209. doi:10.1016/j.jwpe.2015.06.011.

Carliell, C. M., Barclay, S. J., Shaw, C., Wheatley, A. D. & Buckley, C. A. 1998 The effect of salts used in textile dyeing on microbial decolourisation of a reactive azo dye. *Environmental Technology* **19** (11), 1133–1137. doi:10.1080/09593331908616772.

Carvalho, M. G. P., Marcelino, D. M. S., Menezes, O., Foresti, E., Damianovic, M. H. Z., Kato, M. T., Florêncio, L. & Gavazza, S. 2022 The influence of sulphate on the treatment of azo dye-containing wastewater in an anaerobic-microaerobic compartmentalized fixed-bed bioreactor. *The Canadian Journal of Chemical Engineering* **100** (5), 885–892.

Chan, C. & Farahbakhsh, K. 2015 Oxygen demand of fresh and stored sulfide solutions and sulfide-rich constructed wetland effluent. *Water Environment Research* **87** (8), 721–726. doi:10.2175/106143015(14362865225951.

Chen, Y., Cheng, J. J. & Creamer, K. S. 2008 Inhibition of anaerobic digestion process: a review. *Bioresource Technology* **99** (10), 4044–4064. doi:10.1016/j.biortech.2007.01.057.

Chung, K.-T. & Stevens, S. E. J. 1993 Degradation of azo dyes by environmental microorganisms and helminths. *Environmental Toxicology and Chemistry* **12**, 2121–2132. doi:10.1016/0269-7491(92)90127-V.

Cirik, K., Kitis, M. & Cinar, O. 2013 The effect of biological sulfate reduction on anaerobic color removal in anaerobic-aerobic sequencing batch reactors. *Bioprocess and Biosystems Engineering* **36** (5), 579–589. doi:10.1007/s00449-012-0813-2.

Damianovic, M. H. R. Z. & Foresti, E. 2007 Anaerobic degradation of synthetic wastewaters at different levels of sulfate and COD/sulfate ratios in horizontal-flow anaerobic reactors (HAIB). *Environmental Engineering Science* **24** (3), 383–393. doi:10.1089/ees.2006.0067.

Ferraz Jr, A. D. N., Kato, M. T., Florencio, L. & Gavazza, S. 2011 Textile effluent treatment in a UASB reactor followed by submerged aerated biofiltration. *Water Science and Technology* **64** (8), 1581–1589. doi:10.2166/wst.2011.674.

Florêncio, T. M.., Godoi, L. A. G., Rocha, V. C., Oliveira, J. M. S., Motteran, F., Gavazza, S., Vicentine, K. F. D. & Damianovic, M. H. R. Z. 2021 Anaerobic structured-bed reactor for azo dye decolorization in the presence of sulfate ions. *Journal of Chemical Technology & Biotechnology* **96** (6), 1700–1708.

Hao, T. M.., Xiang, P. Y., Mackey, H. R., Chi, K., Lu, H., Chui, H. K., van Loosdrecht, M. C. & Chen, G. H. 2014 A review of biological sulfate conversions in wastewater treatment. *Water Research* **65**, 1–21. doi:10.1016/j.watres.2014.06.043.

Ibrahim, S. M. A. & Attia, S. I. 2015 The influence of the condenser cooling seawater salinity changes on the thermal performance of a nuclear power plant. *Progress in Nuclear Energy* **79**, 115–126. doi:10.1016/j.pnucene.2014.11.004.

Jonstrup, M., Kumar, N., Murto, M. & Mattiasson, B. 2011 Sequential anaerobic-aerobic treatment of azo dyes: decolourisation and amine degradability. *Desalination* **280** (1–3), 339–346. doi:10.1016/j.desal.2011.07.022.

Kato, M. T., Field, J. A. & Lettinga, G. 1997 Anaerobe tolerance to oxygen and the potentials of anaerobic and aerobic cocultures for wastewater treatment. *Brazilian Journal of Chemical Engineering* **14** (4), 395–407. doi:10.1590/S0104-66321997000400015.

Köchling, T., Ferraz Jr, A. D. N., Florencio, L., Kato, M. T. & Gavazza, S. 2016 454-Pyrosequencing analysis of highly adapted azo dye-degrading microbial communities in a two-stage anaerobic–aerobic bioreactor treating textile effluent. *Environmental Technology* **38** (6), 687–693. doi:10.1080/09593330.2016.1208681.

Lens, P. N. L., Visser, A., Janssen, A. J. H., Hulshoff Pol, L. W. & Lettinga, G. 1998 Biotechnological treatment of sulfate-rich wastewaters. *Critical Reviews in Environmental Science and Technology* **28** (1), 41–88. doi:10.1080/10643389891254160.

Levenspiel, O. 1987 *Ingeniería de las reacciones químicas*, 2nd edn. Ediciones Repla City: México, D.F (Mexico City).

Liamleam, W. & Annachhatre, A. P. 2007 Electron donors for biological sulfate reduction. *Biotechnology Advances* **25** (5), 452–463. doi:10.1016/j.biotechadv.2007.05.002.

Maree, J., Gerber, A. & Strydom, W. F. 1986 A biological process for sulphate removal from industrial effluents. *Water SA* **12** (3), 139–144.

Maree, J. P., Hill, E. & Gerber, A. 1987 An integrated process for biological treatment of sulfate-containing industrial. *Journal – Water Pollution Control Federation* **59** (12), 1069–1074.

Menezes, O., Brito, R., Hallwass, F., Florêncio, L., Kato, M. T. & Gavazza, S. 2019 Coupling intermittent micro-aeration to anaerobic digestion improves tetra-azo dye Direct Black 22 treatment in sequencing batch reactors. *Chemical Engineering Research and Design* **146**, 369–378. doi:10.1016/j.cherd.2019.04.020.

Metcalf and Eddy 2003 *Wastewater Engineering: Treatment and Reuse*, 4th edn. McGraw-Hill, New York, USA.

Motteran, F., Gomes, P. C. F. L., Silva, E. L. & Varesche, M. B. A 2017 Simultaneous determination of anionic and nonionic surfactants in commercial laundry wastewater and anaerobic fluidized bed reactor effluent by online column-switching liquid chromatography/tandem mass spectrometry. *Science of the Total Environment* **580**, 1120–1128. doi:10.1016/j.scitotenv.2016.12.068.

Muyzer, G. & Stams, A. J. M. 2008 The ecology and biotechnology of sulphate-reducing bacteria. *Nature Reviews Microbiology* **6** (6), 441–454. doi:10.1038/nrmicro1892.

Oliveira, J. M. S., Silva, M. R. L., Issa, C. G., Corbi, J. J., Damianovic, M. H. R. Z. & Foresti, E. 2020 Intermittent aeration strategy for azo dye biodegradation: a suitable alternative to conventional biological treatments? *Journal of Hazardous Materials* **385**, 121558. doi:10.1016/j.jhazmat.2019.121558.

Pandey, A., Singh, P. & Iyengar, L. 2007 Bacterial decolorization and degradation of azo dyes. *International Biodeterioration and Biodegradation* **59** (2), 73–84. doi:10.1016/j.ibiod.2006.08.006.

Pinheiro, H. M., Touraud, E. & Thomas, O. 2004 Aromatic amines from azo dye reduction: status review with emphasis on direct UV spectrophotometric detection in textile industry wastewaters. *Dyes and Pigments* **61** (2), 121–139. doi:10.1016/j.dyepig.2003.10.009.

Santos, A. B. d., Cervantes, F. J. & van Lier, J. B. 2007 Review paper on current technologies for decolourisation of textile wastewaters: perspectives for anaerobic biotechnology. *Bioresource Technology* **98** (12), 2369–2385. doi:10.1016/j.biortech.2006.11.013.

Silva, M. E. R., Firmino, P. I. M., Sousa, M. R. & dos Santos, A. B. 2012 Sequential anaerobic/aerobic treatment of dye-containing wastewaters: colour and COD removals, and ecotoxicity tests. *Applied Biochemistry and Biotechnology* **166** (4), 1057–1069. doi:10.1007/s12010-011-9493-7.

Singh, K. & Arora, S. 2011 Removal of synthetic textile dyes from wastewaters: a critical review on present treatment technologies. *Critical Reviews in Environmental Science and Technology* **41** (9), 807–878. doi:10.1080/10643380903218376.

Somasiri, W., Li, X. F., Ruan, W. Q. & Jian, C. 2008 Evaluation of the efficacy of upflow anaerobic sludge blanket reactor in removal of colour and reduction of COD in real textile wastewater. *Bioresource Technology* **99** (9), 3692–3699. doi:10.1016/j.biortech.2007.07.024.

Souza, L. F. C., Florencio, L., Gavazza, S. & Kato, M. T. 2016 Methanogenic activity inhibition by increasing the linear alkylbenzene sulfonate (LAS) concentration. *Journal of Environmental Science and Health – Part A Toxic/Hazardous Substances and Environmental Engineering* **51** (8), 656–660. doi:10.1080/10934529.2016.1159876.

Sudha, M., Saranya, A., Selvakumar, G. & Sivakumar, N. 2014 Microbial degradation of azo dyes: a review. *International Journal of Current Microbiology and Applied Sciences* **3** (2), 670–690.

Thauer, R. K., Jungermann, K. & Decker, K. 1977 Energy conservation in chemotrophic anaerobic bacteria. *Bacteriological Reviews* **41** (1), 100–180. doi:10.1108/eb027807.

van der Zee, F. P. & Villaverde, S. 2005 Combined anaerobic-aerobic treatment of azo dyes – a short review of bioreactor studies. *Water Research* **39** (8), 1425–1440. doi:10.1016/j.watres.2005.03.007.

van der Zee, F. P., Lettinga, G. & Field, J. A. 2001 Azo dye decolourisation by anaerobic granular sludge. *Chemosphere* **44** (5), 1169–1176. doi:10.1016/S0045-6535(00)00270-8.

Visser, A. 1995 *The Anaerobic Treatment of Sulfate Containing Wastewater*. PhD Thesis, Wageningen, Netherlands, p. 157.

Woolard, C. R. & Irvine, R. L. 1995 Treatment of hypersaline wastewater in the sequencing batch reactor. *Water Research* **29** (4), 1159–1168. doi:10.1016/0043-1354(94)00239-4.

Yaseen, D. A. & Scholz, M. 2019 Textile dye wastewater characteristics and constituents of synthetic effluents: a critical review. *International Journal of Environmental Science and Technology* **16**, 1193–1226. doi:10.1007/s13762-018-2,130-z.

First received 31 January 2022; accepted in revised form 15 April 2022. Available online 23 April 2022

doi: 10.2166/wst.2022.182

Evaluation of corrosion and scaling potentials of oilfield waters in an offshore producing facility, Niger Delta

T. A. Bolaji [a,*], E. G. Olumayede [b] and A. M. Ojo [b]

a Department of Geology, Federal University Oye-Ekiti, Nigeria
b Department of Industrial Chemistry, Federal University Oye-Ekiti, Oye-Ekiti, Nigeria
*Corresponding author. E-mail: taiwo.bolaji@fuoye.edu.ng

TAB, 0000-0002-5196-1835; EGO, 0000-0002-9554-2261; AMO, 0000-0001-5824-2254

ABSTRACT

In this study, water samples from Miocene reservoirs, offshore Niger Delta, and seawater samples used for water injection were investigated in an attempt to examine the chemistry, evaluate the corrosion behaviour of steel, iron, and aluminium in different aqua media, and evaluate the scaling potentials of the oilfield produced waters (OFPW). Chemical analyses of the waters were determined; corrosion rate measurements were carried out by the weight loss method at room temperature while corrosion kinetics was carried out using conventional methods. Langelier saturation index (LSI), Ryznar stability index (RSI), Larson–Skold index (L–S), Puckorius scaling index (PSI), and aggressiveness index (AI) were evaluated for assessing the corrosiveness and scaling potential of the formation waters, using water quality data. The magnitude of corrosion of these metals was studied for an exposure period of 42 days. Chemical analysis revealed that the waters are slightly alkaline and generally classified as hard, saline water of the Na-Cl type based on its total dissolved solids (TDS). Produced water pH values range from 7.32 to 8.38. Results showed the likelihood of some of the water to form mild to severe scales based on the corrosivity indices, while the seawater samples are classified as 'non-aggressive' and 'aggressive'. Steel has the highest corrosion rate with a value of 3.84×10^{-3} mg cm^{-2} h^{-1} compared to aluminium with the lowest rate of 0.37×10^{-3} mg cm^{-2} h^{-1}. In most cases, the rate of corrosion of the metals followed the first-order rate constant in some of the samples, and the second-order in others within the first seven days. It was observed that the rate of corrosion follows this order: steel > iron > aluminium. The potential heavy and intolerable corrosion associated with the use of these seawater samples as injection waters is a potential risk that must be handled by adequate treatment.

Key words: corrosiveness, kinetics, Niger Delta, produced water, scaling potential

HIGHLIGHTS

- An experimental study of corrosion and scaling potentials of produced water and seawater samples was carried out.
- Water chemistry revealed that the produced waters are saline, dominantly of the Na-Cl type based on its TDS concentrations.
- Five indices: Langelier saturation index (LSI), Ryznar stability index (RSI), Larson–Skold index (L–S index), Puckorius scaling index (PSI), and aggressiveness index (AI) were computed.
- LSI is generally less than zero, in the water samples, indicating supersaturated water with scaling tendencies.
- Corrosion rates of the metals beyond the first week generally followed first-order rate kinetics.

GRAPHICAL ABSTRACT

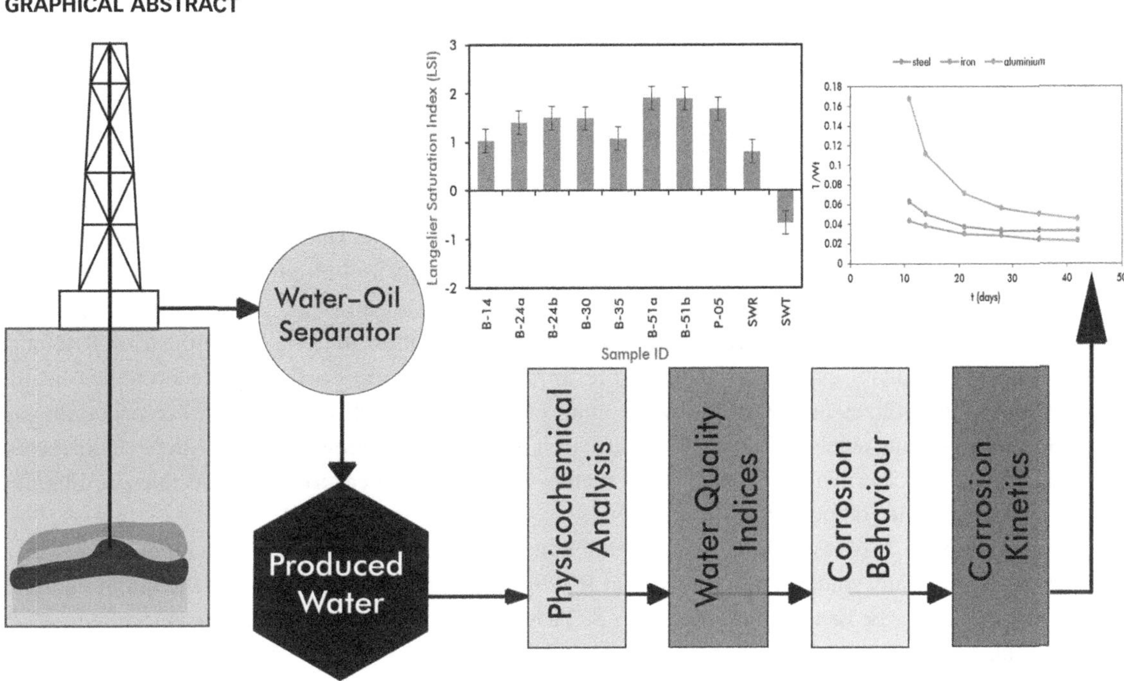

INTRODUCTION

Formation waters occur naturally in association with crude oil and are referred to as oilfield produced water (OFPW) once it is separated from the oil for disposal. Since this water occurs naturally with the hydrocarbon fluids, they tend to generate certain problems in oilfield production operations. The large volume of OFPW, which the oil industry is currently contending with, makes the industry appear more like a water industry since it constitutes the largest waste stream on oilfield production facilities (OFPF) (Du *et al.* 2005). Due to the large volume of OFPW, it is often considered as injection water to enhance hydrocarbon recovery and maintain reservoir pressure, or may also be treated and disposed of into the environment. The former may trigger biogenic souring, corrosion and scaling, while the latter, if not properly treated, may cause bioaccumulation and transfer of potentially toxic elements (PTEs) in aquatic organisms, which would eventually affect the food chain (Konwar *et al.* 2021). OFPW contains low- and high-molecular-weight petroleum hydrocarbons (HCs), dispersed and dissolved oil, dissolved compounds, suspended solids, naturally occurring radioactive minerals (NORM), dissolved gases, and PTEs (Ozgun *et al.* 2013; Kpeglo *et al.* 2016).

Water corrosivity depends on many factors including pH, oxygen content, the presence of metabolizing bacteria, and its suspended solids, which tends to promote microbially assisted corrosion. Scales are inorganic deposits formed due to the precipitation of solids resulting from brines that are present in the oilfield reservoir and production system. The tendency of water to deposit insoluble and protective scales is a function of its corrosivity. Corrosion is the destruction or deterioration of materials due to its reaction with its environment and in an offshore setting, rusting, sweet corrosion, sour and microbiological corrosion are major concerns. Corrosion of metal surfaces is greatly influenced by their chemical composition, electronic properties, microstructure, and passive film properties (Das *et al.* 2009). Water saturated with $CaCO_3$ tends to precipitate scale, which may protect against corrosion, which could also produce encrustation in water and injection wells. The degree of $CaCO_3$ saturation has been used as an indicator of water corrosivity and scaling/encrustation tendency. Metals and their alloys are important materials used in pipeline production for oil fields. Experience and investigation of pipe failures suggest that corrosion of metals, both cast iron and steel, is the most predominant cause of pipe failures (Rajani *et al.* 1995; Mohebbi & Li 2011). In the last two decades, different studies have pointed out to the oil industry the advantages that aluminium alloys may present for tubular manufacturing compared to steel (Gelfgat *et al.* 2005; Osorio-Celestino *et al.* 2020). Since corrosion is linked to almost all pipe failures, it has become a global problem for all stakeholders, in particular engineers and asset managers of buried metal pipes (Goulter 1985; de Sena *et al.* 2012). The impact of corrosion cuts across so many sectors such as oil and gas, car assembling plants, chemical, electric power, medical and engineering industries (Parangusan *et al.* 2021). Nwanonenyi *et al.* (2020) reported the detrimental effects of metal corrosion in industries. Adequate measures that guarantee long-lasting materials must be put in place for efficient service delivery. Different approaches such as the application of native passive films on the surface and other corrosion inhibitory measures are being implemented to minimize corrosion in engineering sites. As is well appreciated, the consequence of pipe failures can be socially, economically, and environmentally catastrophic, resulting in massive disruption of daily life, considerable economic loss, widespread flooding, subsequent environmental pollution and even casualties and so forth (Hou *et al.* 2016). Corrosion generally occurs due to an electrochemical process that occurs in stages. The rates of corrosion also vary with time, depending on a complex interaction between the material, its environment and circumstances of exposure. In as much as the primary factors responsible for corrosion such as water and oxygen are present, the process is inevitable. The estimation is needed for engineers to be able to predict the lifespan of construction materials in industries. The cost of corrosion may run into billions of dollars every year. Due to the impact of corrosion on business economics, health safety and the environment, this study utilizes conventional water analysis, corrosion and scaling indices to ascertain the water stability conditions. Since aluminium, steel and iron are construction materials that show varying degrees of corrosion due to their physicochemical and electrochemical properties, these metals were used to measure corrosion rates in different formation water and seawater media under laboratory conditions.

To manage a potential scaling problem, it is essential to know where and how much scale forms during oil and water production (Osorio-Celestino *et al.* 2020). Many studies have shown that PW has high TDS content, which is the major cause of scale formation and pipe clogging (Arthur & Bruce 2005; Merdhah & Yassin 2007; Al-Ghouti *et al.* 2019; Al-Samhan *et al.* 2020; Bolaji *et al.* 2021). The Niger Delta is a beehive of oil exploration and production (E & P) activities in Nigeria and the volume of OFPW is high. There is scant literature on the effect of OFPW on construction materials used in the area. The present study aims to elucidate the corrosion and scaling potential of these waters on iron, steel and aluminium. This work examines the chemical characteristics of the Freeman OFPW to evaluate the corrosion behaviour of steel, aluminium, and iron in the aqua media, and determine the scaling potentials and corrosion kinetics of the OFPW. This research will provide further information for engineers in ascertaining the integrity and predicting the lifespan of materials used in the construction of OFPF.

GEOLOGICAL SETTING

The study area is located in the production zone of the Freeman field in the southwestern part of the Niger Delta, about 120 km offshore on the continental slope, at water depths of about 950–1,200 m (Figure 1). The reservoirs are structurally and stratigraphically trapped mud-rich unconfined turbidite sands within the Akata-Agbada petroleum system, located in a mid-lower slope depositional setting and are of Lower–Upper Miocene age within the prograding siliciclastic system. The Tertiary section of the Niger Delta is divided into three formations: the Akata, Agbada and Benin Formations, the type sections which are described by Short & Stauble (1967) and summarized in many other papers (Weber & Daukoru 1975; Avbovbo 1978; Evamy *et al.* 1978; Ejedawe 1981; Whiteman 1982; Knox & Omatsola 1989; Doust & Omatsola 1990; Beka & Oti 1995; Bolaji 2020; Bolaji *et al.* 2021; and references cited therein).

MATERIALS AND METHODS

Water sources and experimental method

Nine (9) water samples were extracted (without adding a demulsifier) from produced fluids obtained from the Miocene reservoirs across the Freeman field (Figure 1). Raw (untreated) and treated seawater (injection water) samples were also collected close to the deepwater facility location. Chemical analyses were performed according to Standard Methods for the Examination of Water & Wastewater, SMEWW Standards for inorganic ions, which influence scaling, and corrosion. Corrosion rate measurements were carried out by the weight loss method at room temperature. The experiments were carried out at a room temperature of 29.85 °C (303 °K). A constant volume (70 ml) of the different water samples was placed in well-closed glass bottles and were left at room temperature for 20 min to achieve thermal equilibrium with the surrounding environment. The steel, iron and aluminium specimens were of dimensions $4 \times 4 \times 0.18$ cm. A blank experiment was also conducted by placing the metals in distilled water. The clean, dried specimens were weighed (W_a, g) using a digital balance of 10^{-4} g accuracy and each specimen was inserted carefully in the studied sample waters. The total duration of the experiment was 42 days.

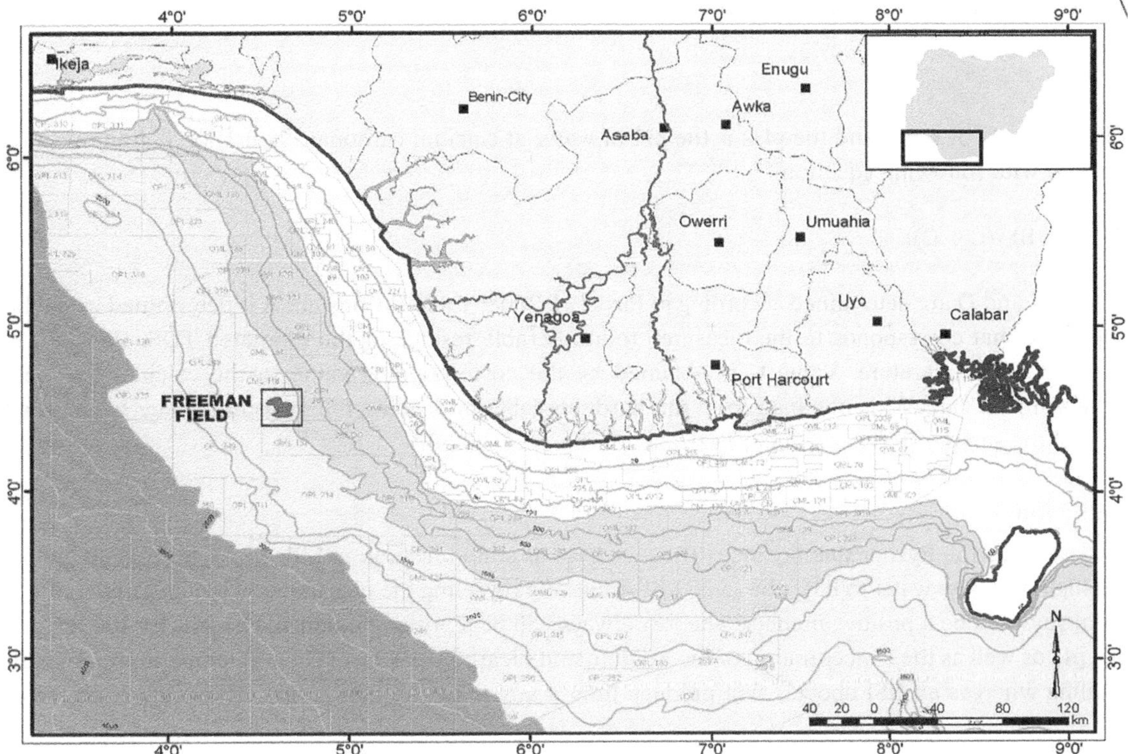

Figure 1 | Niger Delta concession map showing the study area (Bolaji 2020).

The specimens were investigated after 7, 11, 14, 21, 28, 35 and 42 days of exposure. At each time interval, the specimens were removed from the studied solution, washed with double-distilled water, dried and mechanically polished with a 1,200 grade of emery paper, to remove all corrosion products and to obtain a smooth surface. Then the polished specimens were rinsed with ethanol and dried with air and finally weighed (W_b, g). The corrosion rates (CR) (mg cm^{-2} h^{-1}) were calculated according to the following Equation (1):

$$CR = \frac{W_b - W_a}{At} \tag{1}$$

where W_b and W_a are coupon weights (mg) measured before and after immersion in the water samples, A is the exposed area (cm^{-2}) and t (h) is the exposure time. Each experiment was run twice and the average value was taken.

Water stability indices

In this study, five indices were used to identify the corrosiveness and scaling potentials of the formation water and seawater samples. These include Langelier saturation index (LSI), Ryznar stability index (RSI), Larson–Skold index (L–S index), Puckorius scaling index (PSI), and aggressiveness index (AI).

Langelier saturation index

The LSI, developed by Dr Langlier in 1936 is an equilibrium model derived from the theoretical concept of saturation which indicates the degree of saturation of water and its potential to precipitate calcium carbonate in a qualitative assessment (Olajire 2013). LSI considers the effects of calcium, total alkalinity, dissolved solids, and temperature to obtain pHs as regards their tendency to be in equilibrium with calcium carbonate. Calcium carbonate is just one of many minerals found in water and is one of the most important elements responsible for forming calcareous deposits. The LSI considers the effects of calcium, total alkalinity, TDS, and temperature to arrive at a computed pH shown as pHs in the formula. Once the pH is known you simply subtract it from the actual pH of the water and, if the result is positive, the water will be scaling and conversely, if the number is negative, the water will tend to dissolve calcium carbonate (McCaul 2008). After calculation of pHs, the value of LSI is defined as follows (Fazlzadehdavilb *et al.* 2009):

$$LSI = pH - pH_s \tag{2}$$

where, pH is actual pH of water and the pHs is the pH of water at calcium carbonate saturation condition (McCaul 2008). pHs is calculated with following equation:

$$pH_s = [(9.5 + A + B) - (C + D)] \tag{3}$$

Values of A, B, C and D are determined according to Piri *et al.* 2008; Table 1. Constant A is determined using the TDS table by taking the value that corresponds to the measured total filterable residue or the estimated TDS. Constant B takes into account the effect of temperature. Value C is obtained by the corresponding value to the calcium hardness (in mg/L $CaCO_3$) of the sample. Value D is obtained from the hardness table by reading the measured value for total alkalinity (in mg/L $CaCO_3$) of the sample (Piri *et al.* 2008; Fazlzadehdavilb *et al.* 2009).

Ryznar stability index

John Ryznar developed the Ryznar stability index (RSI) in 1944 as a modification to LSI, when he realized the possibility for both low and high hardness water to have the same LSI values. By reversing the placement of pH and pH_s in the formula, the result of RSI always will be a positive number. The RSI always will be a positive number. The pH_s for the RSI is determined via the actual pH as well as the concentration of the calcium and bicarbonate ions, TDS and temperature. An RSI less than 5 should be scaling whereas an RSI above 7 will produce little if any scale (McCaul 2008). The value of RSI is evaluated as follows:

$$RSI = 2(pH_s) - pH \tag{4}$$

where, pH is the measured water pH, and pH_s is the pH at saturation in calcite or calcium carbonate.

Larson–Skold index

The LSI index describes the corrosivity of water towards mild steel. The index is based upon an evaluation of *in situ* corrosion of mild steel. The index is the ratio of equivalents per million (epm) of sulphate (SO_4^{2-}) and (Cl^-) to the epm of alkalinity in the form of bicarbonate plus carbonate.

$$L-S \; Index \; = \frac{[(epm \; Cl^-) \; + \; (epm \; SO_4^{2-})]}{[(epm \; HCO_3^-) \; + \; (epm \; CO_3^{2-})]} \tag{5}$$

This index has been correlated to observed corrosion rates and the type of attack. The L–S index might be interpreted by the following guidelines: index < <0.8, chlorides and sulphates probably will not interfere with natural film formation 0.8 < < index < <1.2, chlorides and sulphates may interfere with natural film forming. Higher than desired corrosion rates might be anticipated. Index > >1.2, the tendency towards high corrosion rates of a local type should be expected as the index increases.

Puckorius scaling index

Paul Puckorius developed the PSI which accounts for two additional variables that the indices do not – the buffering capacity of water and the maximum quantity of precipitate that brings water to equilibrium. Therefore, the PSI uses an equilibrium pH to account for the buffering capacity of water and the maximum quantity of precipitate that brings water to equilibrium. Therefore, the PSI uses an equilibrium PH rather than the actual pH to account for the buffering effect (McCaul 2008). The equilibrium pH is determined as:

$$pHeq = 1.465 + \log (T.ALK) + 4.54 \tag{6}$$

Table 1 | Summary of the computed indices for the studied water samples

Sample ID	pH	TDS	Temp (°C)	Ca^{2+}	Mg^{2+}	$Na^+ + K^+$	T.Alk	A	B	C	D	pH_s
B-01	7.68	22,246.00	25	545.09	349.92	13,389.13	1,160.00	0.335	2.088	2.336	3.064	6.322
B-14	7.32	23,984.00	25	657.32	233.28	7,797.67	1,040.00	0.338	2.088	2.418	3.017	6.291
B-24a	7.73	14,617.00	25	416.83	330.48	3,787.87	1,440.00	0.316	2.088	2.220	3.158	6.326
B-24b	7.75	14,537.00	25	601.20	121.50	3,619.80	1,200.00	0.316	2.088	2.379	3.079	6.246
B-30	8.00	31,761.00	25	561.12	1,117.80	10,907.67	760.00	0.350	2.088	2.349	2.881	6.509
B-35	7.60	23,568.00	25	432.86	806.76	8,021.33	920.00	0.337	2.088	2.236	2.964	6.525
B-51a	7.93	28,430.00	25	737.47	330.48	7,673.60	1,760.00	0.345	2.088	2.468	3.246	6.020
B-51b	8.38	17,671.43	25	368.74	797.04	4,504.00	1,160.00	0.325	2.088	2.167	3.064	6.482
P-05	7.88	24,043.00	25	689.38	408.24	7,784.07	1,240.00	0.338	2.088	2.438	3.093	6.195
SW_R	7.76	36,184.00	25	426.45	1,240.27	10,261.00	360.00	0.356	2.088	2.230	2.556	6.958
SW_T	6.63	34,658.00	25	391.18	1,281.10	9,174.00	178.00	0.354	2.088	2.192	2.250	7.299

Sample ID	LSI	RSI	Cl^-	SO_4^{2-}	CO_3^{2-}	HCO_3^-	L–S Index	pH_{eq}	PSI	AI	T. Hardness
B-01	1.36	4.964	12,043.30	600.00	696.00	1,415.20	5.989	9.069	3.575	13.481	2800.00
B-14	1.03	5.263	12,956.49	1,100.00	624.00	1,268.80	7.426	9.022	3.561	13.155	2600.00
B-24a	1.40	4.923	7,693.72	400.00	864.00	1,756.80	3.088	9.163	3.490	13.508	2400.00
B-24b	1.50	4.743	7,211.32	500.00	720.00	1,464.00	3.531	9.084	3.408	13.608	2000.00
B-30	1.49	5.017	18,761.21	2,000.00	456.00	927.20	15.010	8.886	4.131	13.630	6000.00
B-35	1.07	5.451	13,313.82	1,000.00	552.00	1,122.40	8.349	8.969	4.082	13.200	4400.00
B-51a	1.91	4.111	15,672.24	1,300.00	1,056.00	2,147.20	5.299	9.251	2.790	14.043	3200.00
B-51b	1.90	4.584	9,712.67	400.00	696.00	1,415.20	4.790	9.069	3.894	14.011	4200.00
P-05	1.69	4.509	13,480.58	600.00	744.00	1,512.80	6.239	9.098	3.291	13.812	3400.00
SW_R	0.80	6.156	21,071.83	2,450.00	216.00	439.20	35.900	8.561	5.355	12.946	6168.00
SW_T	−0.67	7.969	21,278.29	2,300.00	106.80	217.16	72.781	8.255	6.344	11.473	6248.00

The numbers resulting from Equation (6) are the same as the RSI so a value less than 5 will be scaling and a number greater than 7 will result in little if any scaling. PSI is calculated as follows (McCaul 2008):

$$PSI = 2(pH_s) - pH_{eq} \tag{7}$$

Aggressiveness index (AI)

The AI was developed for asbestos cement pipes with water temperatures ranging from 4 to 27 °C (40–80 °F). The AI is calculated as a function of pH, calcium concentration, and alkalinity. Waters with AI less than 10 are considered highly aggressive; AI between 10 and 12 is considered mildly aggressive; AI greater than 12 is considered non-corrosive and depositing (McCaul 2008). The AI is calculated as follows (Fazlzadehdavilb *et al.* 2009):

$$AI = pH + \log[(A)(H)] \tag{8}$$

where A is the total alkalinity in mg/L as $CaCO_3$ and H is the calcium hardness (in mg/L $CaCO_3$).

RESULTS AND DISCUSSION

Physicochemical characteristics

Analytical results of the measured physicochemical parameters are presented in Table 2. Produced water pH values range from 7.32 to 8.38, slightly alkaline to alkaline in nature, while those obtained for the raw seawater (SW_R) and treated seawater (SW_T) are 7.76 and 6.63 respectively. pH value less than 4 encourages corrosion as protective oxide film dissolves, while solutions with pH values between 4 and 10 have little effect on corrosion. Higher values above 10 make the metals passive, thereby reducing corrosion. Total alkalinity (T. Alk) ranges between 760 and 1,760 mg/L for formation water samples, while 178 and 360 mg/L were obtained for SW_R and SW_T samples respectively. The mean Total Hardness (TH) concentration of 3,444.44 mg/L for the formation water samples indicate the presence of more magnesium rather than calcium, while TH in seawater samples slightly exceeds the peak value of 6,000 mg/L obtained for the formation waters. Chloride,

Table 2 | Water chemistry results for the studied water samples

S/N	Parameter	Unit	Min.	Max.	Mean	Median	Std Dev.
1	pH	°C	6.63	8.35	7.70	7.75	0.44
2	Eh	*mV*	−0.07	0.04	−0.03	−0.04	0.03
3	EC	mS/cm	19.39	59.55	35.75	32.98	13.65
4	Ca^{2+}	*mg/L*	368.74	737.47	529.79	545.09	130.06
5	Mg^{2+}	*mg/L*	121.50	1,281.10	637.90	408.24	426.12
6	Na^+	*mg/L*	3,576.33	13,328.00	7,811.06	7,742.67	3,026.14
7	K^+	*mg/L*	38.93	194.00	90.77	61.13	53.35
8	Fe	*mg/L*	5.73	268.00	53.37	12.93	94.15
9	NH_4^+	*mg/L*	0.60	3.00	1.79	1.80	0.93
10	Cl^-	*mg/L*	7,211.32	21,278.29	13,926.86	13,313.82	4,881.45
11	SO_4^{2-}	*mg/L*	400.00	2,450.00	1,150.00	1,000.00	770.39
12	HCO_3^-	*mg/L*	217.16	2,147.20	1,244.18	1,415.20	554.61
13	NO_3^-	*mg/L*	6.10	19.10	12.91	12.70	5.24
14	TH	*mg/L*	2,000.00	6,248.00	3,946.91	3,400.00	1,577.43
15	TA	*mg/L*	178.00	1,760.00	1,019.82	1,160.00	454.60
16	TDS	*mg/L*	14,537.00	36,184.00	24,699.95	23,984.00	7,466.03
17	Salinity	‰	13.03	38.44	25.16	24.05	8.82

bicarbonate, and sulphate are the dominant anions, with an average of 12, 316.15 mg/L, 1,447.73 mg/L, and 877.78 mg/L respectively. Sodium and calcium on the other hand predominate the cations, averaging 7,426.79 mg/L and 556.67 mg/L respectively. Total dissolved solids (TDS) for the produced water samples ranges between 14,537 and 31,761 mg/L (*Av.* 22,317.49 mg/L), while 36,184 mg/L and 34,658 mg/L were obtained for the SW_R and SW_T samples. The analyses

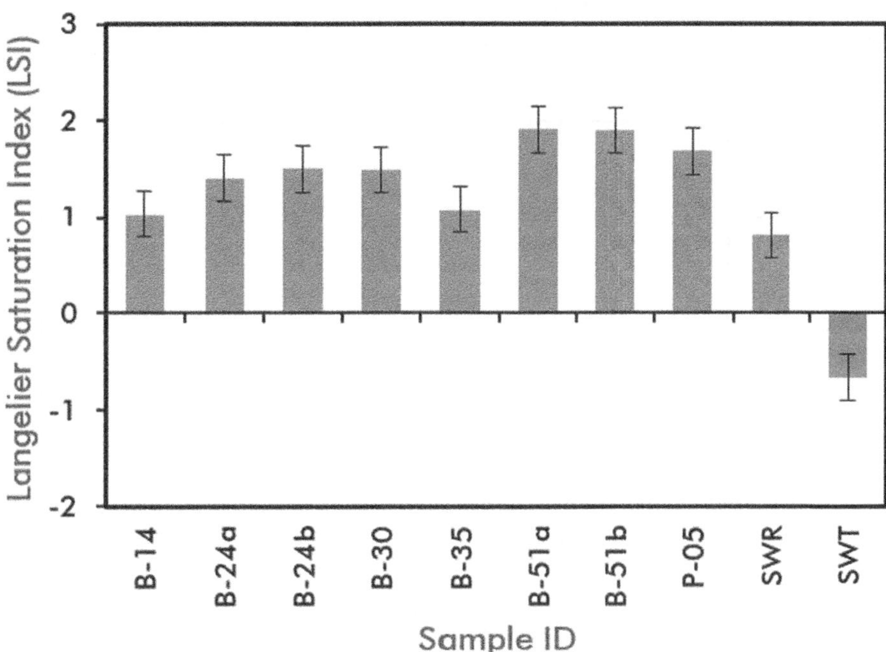

Figure 2 | Computed LSI for the studied water samples.

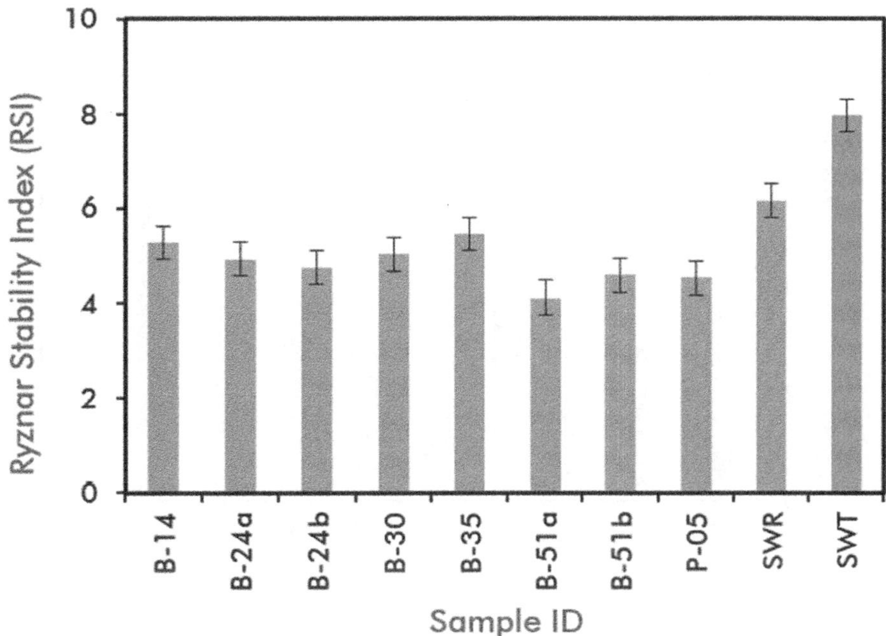

Figure 3 | RSI computed for the studied water samples.

revealed that the principal hydrochemical is Na-K-SO$_4$-Cl where the chemical properties of the waters are dominated by alkali elements. The water chemistry revealed that the formation waters are generally hard, saline water, dominantly of the Na-Cl type based on its TDS concentrations (Sawyer *et al.* 1994; Bolaji *et al.* 2021).

Water quality indices

The LSI determined for the produced water samples ranged from -0.67 to 1.91. Except for the treated seawater (SW$_T$) sample, LSI was generally greater than zero, indicating super saturated water with scaling tendencies (Figure 2). Based on this index, SW$_T$ is not saturated and has no corroding tendency, hence, it is considered suitable as injection water. However,

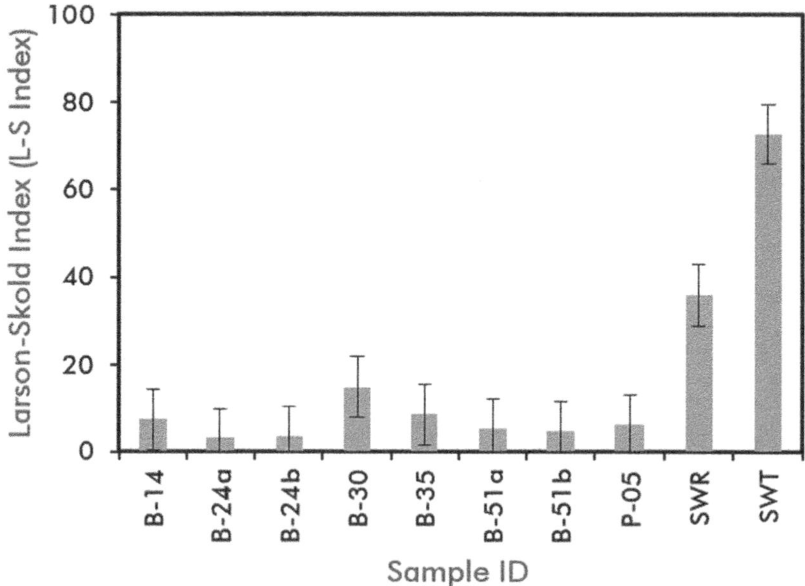

Figure 4 | L–S index computed for the studied water samples.

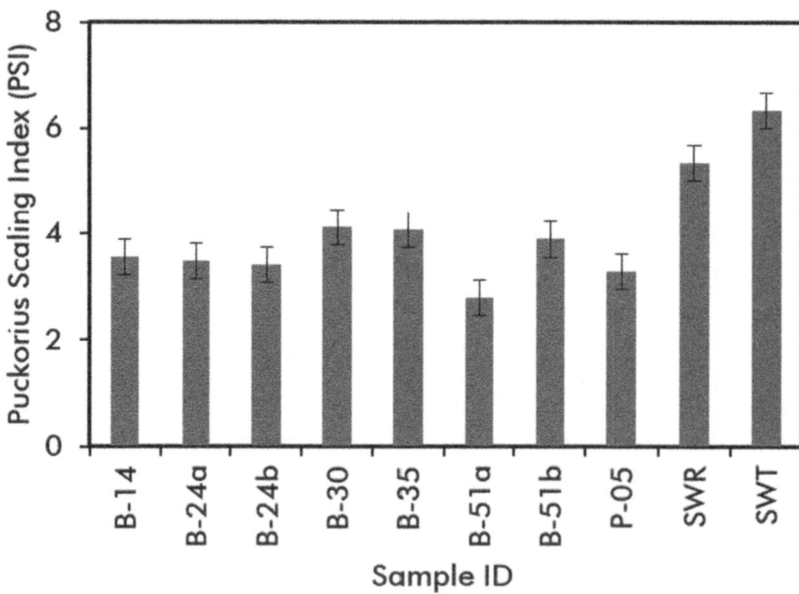

Figure 5 | Computed PSI for the studied water samples.

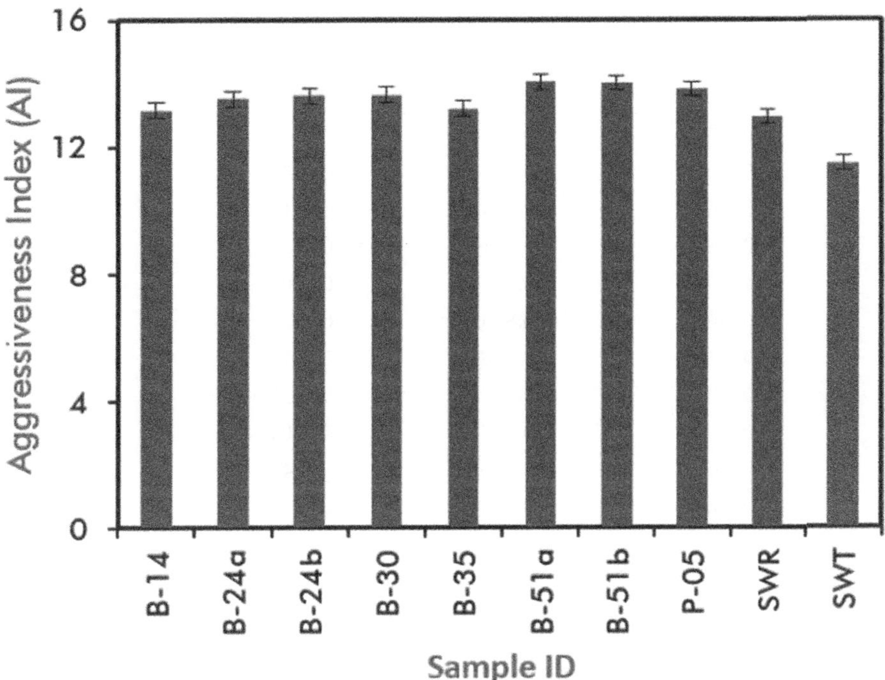

Figure 6 | AI Computed for the studied water samples.

Table 3 | Cumulative corrosion rate (mg·cm^{-2} h^{-1})

S/N	Sample ID	CR steel $\times$ 10^{-3}	CR iron $\times$ 10^{-3}	CR aluminium $\times$ 10^{-3}
1	B-01	3.16	1.30	0.68
2	B-14	2.60	2.11	0.74
3	B-24a	1.92	2.48	0.37
4	B-25b	3.84	2.36	0.49
5	B-30	2.48	3.29	0.99
6	B-35	2.23	3.28	0.99
7	B-51a	1.49	2.11	0.74
8	B-51b	1.79	1.98	1.05
9	P-05	1.86	2.67	1.36
10	SW$_R$	2.66	2.73	2.48
11	SW$_T$	2.54	1.24	0.74

the remaining PW samples are generally supersaturated with respect to $CaCO_3$ and may form scale. Although water temperature has a positive effect on LSI, i.e. as water temperature increases, LSI becomes more positive. In this study, LSI values revealed that the majority of the PW are supersaturated with respect to $CaCO_3$ and scale forming may occur on the pipelines, thereby restricting the free flow of HC fluids. Undersaturated waters (e.g. SW$_T$) changes the chemical equilibrium of $CaCO_3$, favouring the dissolution of protective coatings in pipelines and equipment (DeMartini 1938).

Ryznar saturation index ranges between 4.1 and 5.5 for PW samples and 6.2–8.0 for SW samples (Figure 3). RSI less than 5.5 indicate heavy scale formation while values greater than 8.5 indicates that the water is aggressive, that is such water

Table 4 | First-order and second-order kinetics data

Sample ID	Steel		Iron		Aluminium	
	First order (r^2)	Second order (r^2)	First order (r^2)	Second order (r^2)	First order (r^2)	Second order (r^2)
B-01	0.6543	0.5674	0.3826	0.3549	0.7413	0.6814
B-14	0.5835	0.5043	0.7665	0.6851	0.9572	0.9473
B-24a	0.4474	0.307	0.5044	0.3913	0.1541	0.808
B-24b	0.5156	0.4594	0.5819	0.4887	0.9088	0.798
B-30	0.7181	0.6731	0.6514	0.5209	0.6826	0.6144
B-35	0.6237	0.5691	0.8252	0.7827	0.8829	0.8211
B-51a	0.716	0.5393	0.748	0.6713	0.7606	0.7216
B-51b	0.4681	0.5417	0.639	0.5417	0.767	0.6194
P-05	0.742	0.7111	0.9366	0.9004	0.8717	0.7564
SW_R	0.5505	0.543	0.6079	0.5243	0.7546	0.7197
SW_T	0.4425	0.366	0.565	0.4038	0.6484	0.5197

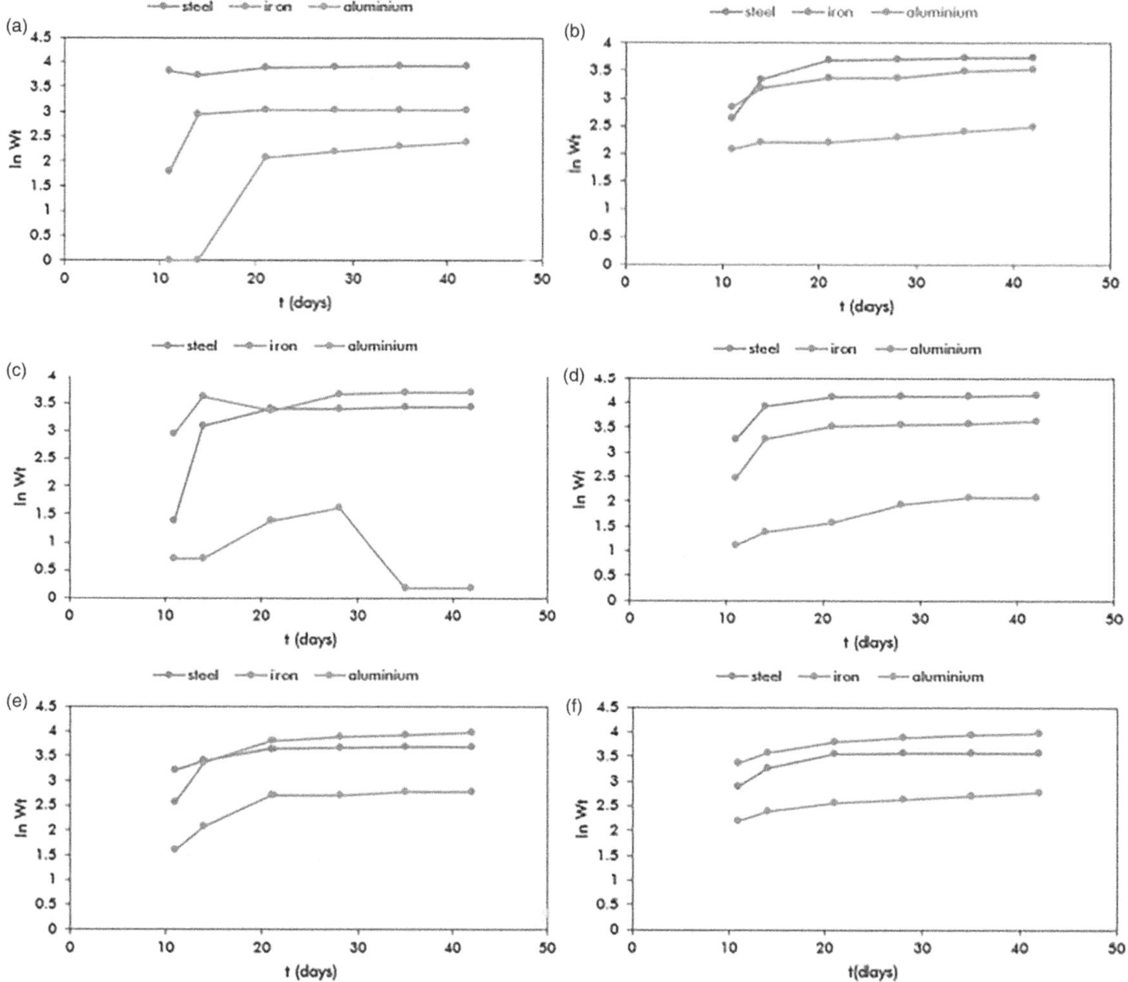

Figure 7 | First-order corrosion kinetic for the studied water samples (a) B-01, (b) B-14, (c) B-24a, (d) B-24b, (e) B-30, and (f) B-35.

actively seeks to gain minerals and metals which results in the corrosion of iron and copper surfaces and causes dissolving concrete and stone surfaces (Konwar *et al.* 2021). The Freeman PW samples show rigorous scaling tendency based on the RSI. SW_R is balanced with no scaling and corrosive tendency, while SW_T has a corrosive tendency. The SW_T which show a seemingly neutral has the highest RSI value.

L–S index is generally greater than 1.2 indicating that high rates of localized corrosion can be expected in all the studied water samples (Figure 4). PSI values in the OFPW samples range from 2.79 to 6.34. PSI values generally suggest that scaling is likely to occur in the PW and SW_R samples, while SW_T shows little scaling and corrosive tendency (Figure 5). AI range from 11.47 to 14.04 (Figure 6). According to the AI, produced water and SW_R samples are non-aggressive with the ability for scale formation, while SW_T is moderately corrosive and aggressive.

Corrosion rates

Steel has the highest CR in B-25b with value 3.84×10^{-3} mg cm^{-2} h^{-1}, while the least value was 1.49×10^{-3} mg cm^{-2} h^{-1} in B-51a (Table 3). Iron has the highest CR in B-30 with a value of 3.29×10^{-3} mg cm^{-2} h^{-1} while the least value was

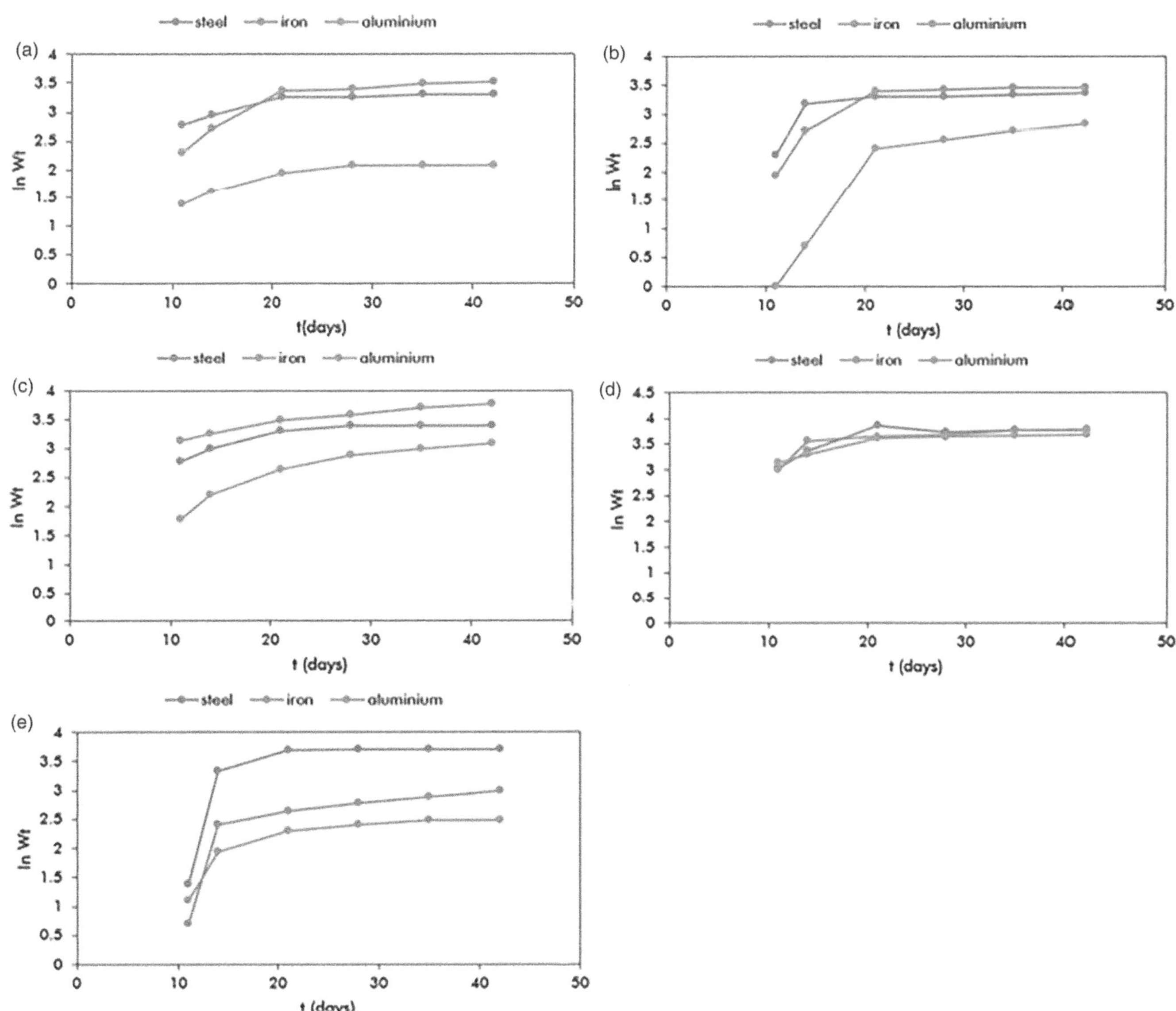

Figure 8 | First-order corrosion kinetics for the studied water samples: (a) B-51a, (b) B-51b, (c) P-05, (d) SW_R, and (e) SW_T.

1.24×10^{-3} mg cm^{-2} h^{-1} in SW$_T$. The highest and least CR were 2.48×10^{-3} mg cm^{-2} h^{-1} in SW$_R$ and 0.37×10^{-3} mg cm^{-2} h^{-1} in B-24a for aluminium. The highest CR was observed in steel with a value of 3.84×10^{-3} mg cm^{-2} h^{-1}, while the least CR occurred in aluminium with a value of 0.37×10^{-3} mg cm^{-2} h^{-1}. The corrosion suffered by iron plates is of the general type (uniform attack). The corrosion rate of iron varied from month to month. Iron has the highest CR in all the water samples except B-25b, B-14, B-01 and SW$_T$ where steel CR was higher. The corrosion suffered by steel plates was mainly of a general type attack. Aluminium has the least CR in all the water samples compared to other metals investigated. It is a very reactive metal but mild to corrosion. Of all metals investigated by Natesan *et al.* (2006), aluminium was also the least corroded. No significant attack was observed on aluminium panels. Corrosion observed in B-01 was due to atmospheric gases and water which aids the corrosion process. An increase in salt content (salinity) also increases the rate of corrosion.

Kinetics of steel, iron and aluminium corrosion in water samples

The kinetics study was investigated at 29.85 °C (303 °K) by fitting the corrosion data into different rate laws (Table 4). Coefficients of determination (R^2) were used to determine the best rate law for the corrosion process. The rate laws considered

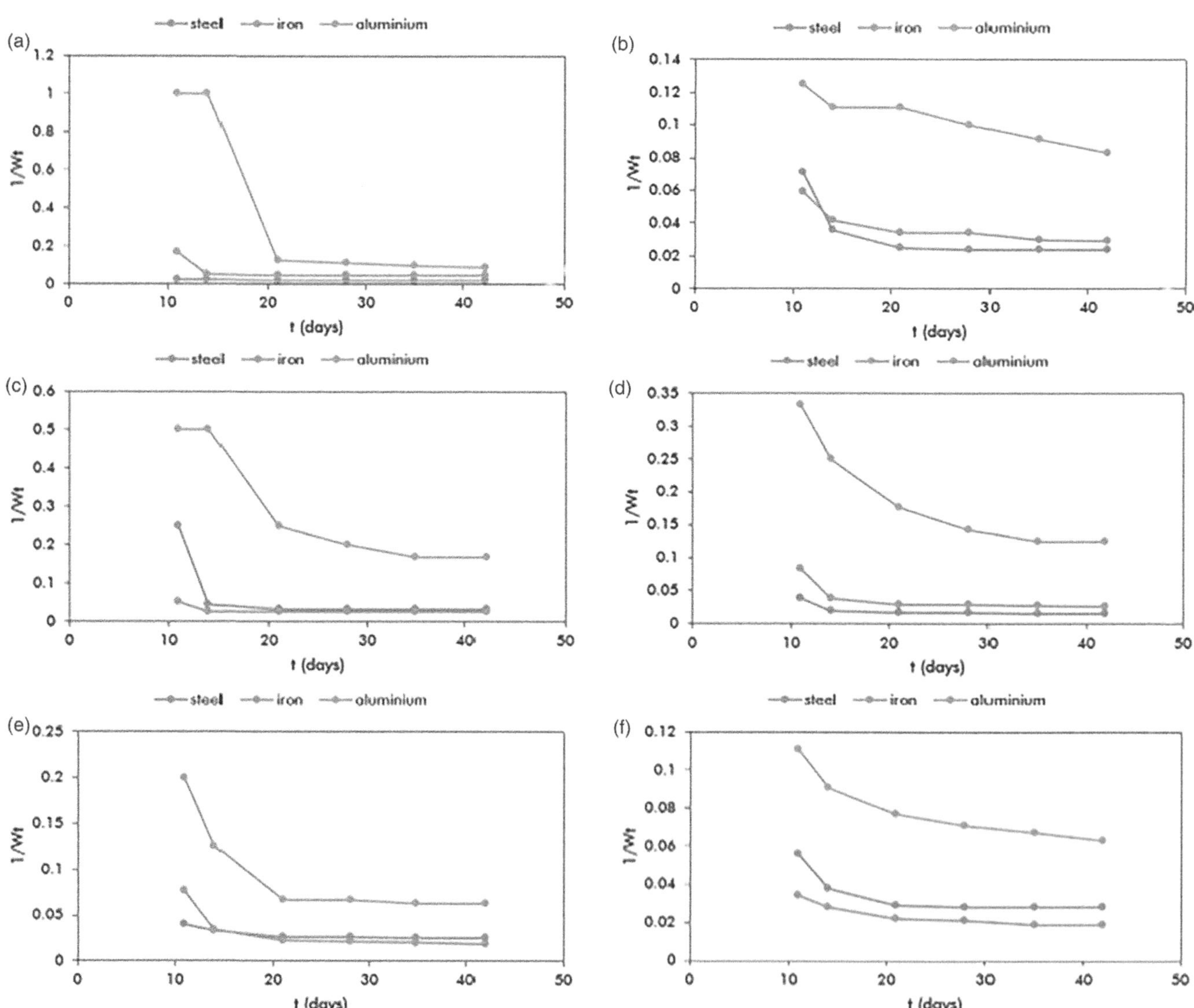

Figure 9 | Second-order corrosion kinetics for the studied water samples (a) B-01, (b) B-14, (c) B-24a, (d) B-24b, (e) B-30, and (f) B-35.

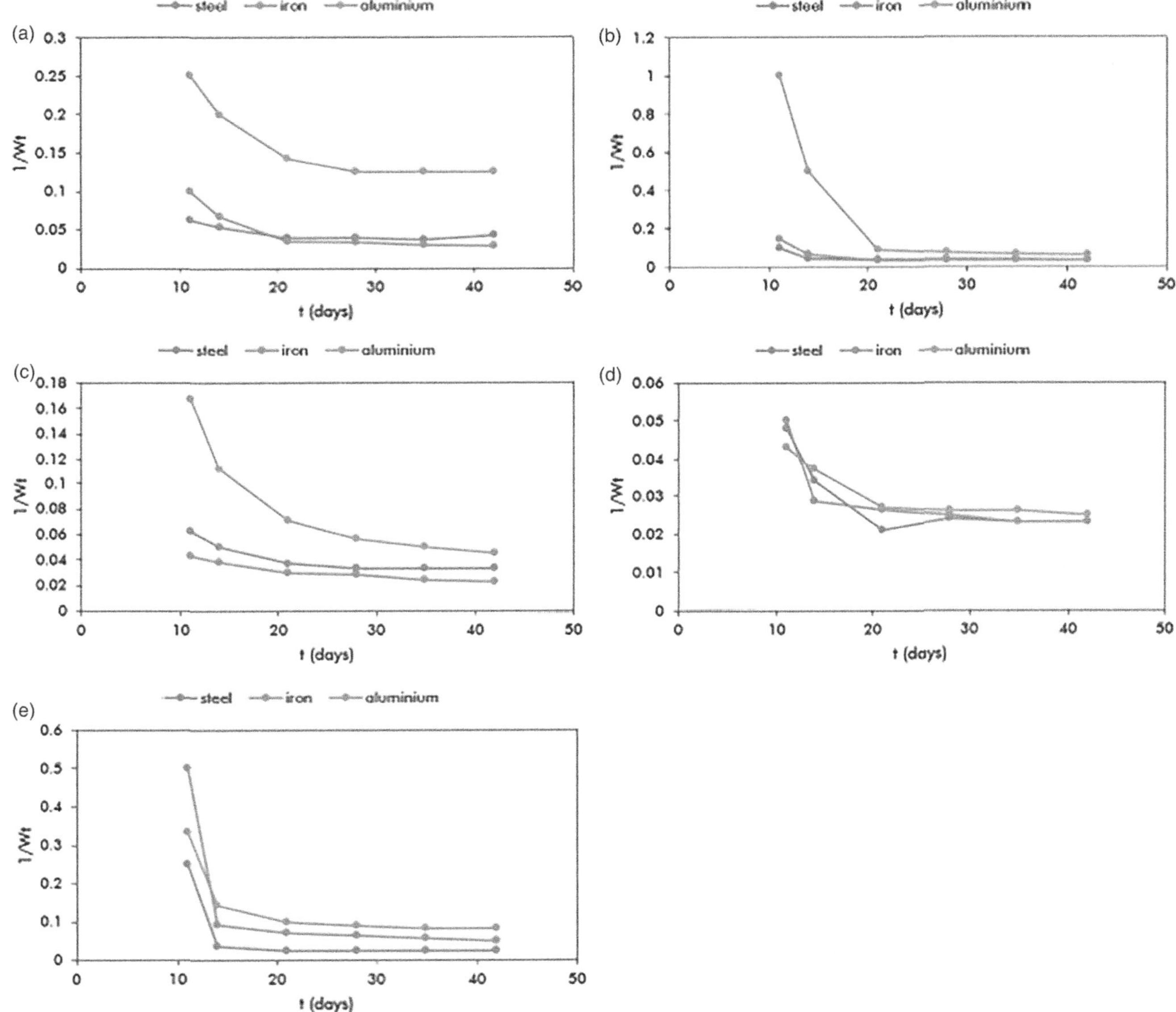

Figure 10 | Second-order corrosion kinetics for the studied water samples (a) B-51a, (b) B-51b, (c) P-05, (d) SW_R, and (e) SW_T.

were:

First order: $\ln W_t = -Kt + \ln W_o$ (9)

Second order: $1/W_t = Kt + 1/W_o$ (10)

where W_o is the initial weight of iron, W_t is the weight loss of iron at time t and k is the rate constant. Previous studies have shown that the rate of corrosion is not constant with respect to time and this can be attributed to complex reaction chemistry, the reactivity of the corrosive medium, adsorption, coating on the metal surface and variations in environmental conditions. Based on results obtained for the first-order (Figures 7 and 8) and second-order (Figures 9 and 10) kinetics, the model with the highest R^2 was chosen as the best fit for the study. By far, most of the results fit into first-order kinetics. The plot of $\ln W_t$ against t was linear and shows a higher R^2 value for all water samples used for steel corrosion study except B-51b. The plot of $1/W_t$ against t for steel in B-51b has a higher R^2 value confirming the second-order reaction model. This result is similar to a study conducted by Aljbour (2016), which indicates a second-order reaction model for steel in various concentrations of HCl solutions. The highest R^2 value confirms the first-order reaction model for iron in all the water samples investigated. Research conducted by Patil & Sharma (2011) also reveals a first-order reaction model for iron in different concentrations of HCl solutions. In almost all the water samples, aluminium fits into the first-order reaction model with R^2 close or higher than 0.9, the only exception is for the B-24b water sample showing a higher R^2 value confirming the second-order kinetic model.

CONCLUSIONS

Chemical analysis revealed that the studied formation water and seawater samples are slightly alkaline and generally classified as hard, saline water based on its TDS. The waters are of the Na-Cl type with Na-K-SO$_4$-Cl as the principal hydrochemical facies. Based on the computation of the different corrosivity indices, there is a likelihood of some of the water to form mild to severe scales, while the seawater samples are classified as 'non-aggressive' and 'aggressive'. In most cases, the rate of corrosion of the metals followed the first-order rate constant in some of the samples, and second-order in others within the first seven days. It was observed that the rate of corrosion follows this order: steel > iron > aluminium. The potential heavy and intolerable corrosion associated with the use of these seawater samples as injection waters is a potential risk that must be handled by adequate treatment. A comprehensive analysis of steel, iron and aluminium corrosion in oilfield waters was presented in this paper. From the results available it has been established that aluminium was the least corroded while the iron was the most corroded. It can be concluded that the results of cumulative corrosion rate and kinetics study presented in the paper can contribute to the body of knowledge of corrosion behaviour of the metals and aid in the prediction of long-term properties of the metals in construction. This knowledge can enable a more accurate prediction of failures of metal pipes.

ACKNOWLEDGEMENTS

The authors wish to appreciate Shell Nigeria for providing fluid samples for this study. Comments from anonymous reviewers, which greatly improved the quality of this work are gratefully acknowledged.

DATA AVAILABILITY STATEMENT

All relevant data are included in the paper or its Supplementary Information.

CONFLICT OF INTEREST STATEMENT

The authors declare there is no conflict.

REFERENCES

Al-Ghouti, M. A., Al-Kaabi, M., Ashfaq, M. & Da'na, D. A. 2019 Produced water characteristics, treatment and re-use: a review. *Journal of Water Process Engineering* **28**, 222–239. https://doi.org/10.1016/j.jwpe.2019.02.001.

Aljbour, S. H. 2016 Modeling of corrosion kinetics of mild steel in hydrochloric acid in the presence and absence of a drug inhibitor. *Portugaliae Electrochimica Acta* **34** (6), 407–416. http://dx.doi.org/10.4152/pea.201606407.

Al-Samhan, A., Alanezi, K., Al-Fadhli, J., Al-Attar, F., Mukadam, S. & George, J. 2020 Evaluating scale deposition and scale tendency of effluent water mix with seawater for compatible injection water. *Journal of Petroleum Exploration and Production Technology* **10**, 2105–2111. https://doi.org/10.1007/s13202-020-00849-w.

Arthur, J. & Bruce, G. 2005 *Technical Summary of oil and gas Produced Water Treatment Technologies*. All Consulting, LLC, Tulsa.

Avbovbo, A. A. 1978 Tertiary lithostratigraphy of the Niger delta. *AAPG Bulletin* **62**, 295–300.

Beka, F. T. & Oti, M. N. 1995 The Distal Offshore Niger Delta: frontier prospects of a mature petroleum province. In: *Geology of Deltas* (Oti, M. N. & Postma, G., eds). A.A. Balkema, Rotterdam, pp. 237–241.

Bolaji, T. A. 2020 *Reservoir Souring Possibilities in Freeman Oilfield, Niger Delta: Insights from Mineralogy, Diagenesis and Water Chemistry*. Unpublished Ph.D. Dissertation, University of Port Harcourt, Nigeria.

Bolaji, T. A., Oti, M. N., Onyekonwu, M. O., Bamidele, T., Osuagwu, M., Chiejina, L. & Elendu, P. 2021 Preliminary geochemical characterization of saline formation water from Miocene Reservoirs, Offshore, Niger Delta. *Heliyon* **7** (2), e06281. https://doi.org/10.1016/j.heliyon.2021.e06281.

Das, N. K., Suzuki, K., Ogawa, K. & Shoji, T. 2009 Early stage SCC initiation analysis of FCC Fe–Cr–Ni ternary alloy at 288 °C: a quantum chemical molecular dynamics approach. *Corrosion Science* **51** (4), 908–913. https://doi.org/10.1016/j.corsci.2009.01.005.

DeMartini, F. E. 1938 Corrosion and the Langelier calcium carbonate saturation index. *Journal (American Water Works Association)* **30** (1), 85–111.

de Sena, R. A., Bastos, I. N. & Platt, G. M. 2012 Theoretical and experimental aspects of the corrosivity of simulated soil solutions. *ISRN Chemical Engineering* **2012**. Article ID 103715. https://doi.org/10.5402/2012/103715.

Doust, H. & Omatsola, E. 1990 Niger Delta in divergent/passive margin basins. In: *AAPG Memoir 48* (Edwards, J. D. & Santogrossi, P. A., eds). American Association of Petroleum Geologists, Tulsa, pp. 201–238.

Du, Y., Guan, L. & Liang, H. 2005 Advances of produced water management. In *Canadian International Petroleum Conference*, Petroleum Society of Canada. https://doi.org/10.2118/2005-060.

Ejedawe, J. E. 1981 Patterns of incidence of oil reserves in Niger Delta Basin. *American Association of Petroleum Geologists Bulletin* **65**, 1574–1585.

Evamy, B. D., Haremboure, J., Kamerling, P., Knapp, W., Molloy, F. A. & Rowlands, P. H. 1978 Hydrocarbon habitat of Tertiary Niger Delta. *American Association of Petroleum Geologists Bulletin* **62**, 1–39.

Fazlzadehdavilb, D. M., Norouzi, M., Mazloomi, S., Amarluie, A., Tardast, A. & Karamitabar, Y. 2009 Survey of corrosion and scaling potentials of produced water from Ilam water treatment plant. *World Applied Sciences Journal* **7** (Special Issue of Applied Math).

Gelfgat, M. Y., Basovich, V. S. & Adelman, A. J. 2005 Aluminium alloy tubulatrs for oil and gas industry. In *SPE Annual Technical Conference and Exhibition*, Society of Petroleum Engineers, Dallas, Texas, pp. 1–11.

Goulter, C. 1985 An analysis of pipe breakage in urban water distribution networks. *Canadian Journal of Civil Engineering* **12** (2), 286–293.

Hou, Y., Lei, D., Li, S., Yang, W. & Li, C. 2016 Experimental investigation on corrosion effect on mechanical properties of buried metal pipes. *International Journal of Corrosion*. https://doi.org/10.1155/2016/5808372.

Knox, G. J. & Omatsola, E. M. 1989 Development of the Cenozoic Niger Delta in terms of the 'Escalator Regression' model and impact on hydrocarbon distribution. In: *Proceedings of the KNGMG Symposium Coastal Lowlands, Geology and Geotechnology, The Hague, 1987* (van der Linden, W. J. M., Cloetingh, S. A. P. L., Kaasschieter, J. P. K., van der Graf, W. J. E., Vandenberglie, J. & van der Gun, J. A. M., eds). Kluwer Academic, Dordrecht, pp. 181–202.

Konwar, D., Gogoil, S. B. & Gogoi, T. J. 2021 Evaluation of the corrosion and scaling potentials of oilfield produced water of the Upper Assam Basin. In: *Advances in Petroleum Technology* (Gogoi, S. B., ed.). Jenny Stanford Publishing Pte. Ltd., Singapore, p. 505.

Kpeglo, D. O., Mantero, J., Darko, E. O., Emi-Reynolds, G., Faanu, A., Manjón, G., Vioque, I., Akaho, E. H. K. & Garcia-Tenorio, R. 2016 Radiochemical characterization of produced water from two production offshore oilfields in Ghana. *Journal of Environmental Radioactivity* **152**, 35–45. http://dx.doi.org/10.1016/j.jenvrad.2015.10.026.

McCaul, C. 2008 Stress corrosion cracking. *Materials Newsletter* **7** (4).

Merdhah, A. & Yassin, A. 2007 Scale formation in oil reservoir during water injection of high-salinity formation water. *Journal of Applied Sciences* **7** (21), 3198–3207. https://dx.doi.org/10.3923/jas.2007.3198.3207.

Mohebbi, H. & Li, C. Q. 2011 Experimental investigation on corrosion of cast iron pipes. *International Journal of Corrosion* **2011**. Article ID506501. https://doi.org/10.1155/2011/506501.

Natesan, M., Venkatachari, G. & Palaniswamy, N. 2006 Kinetics of atmospheric corrosion of mild steel, zinc, galvanised iron and aluminium at 10 exposure stations in India. *Corrosion Science* **48**, 3584–3608. http://dx.doi.org/10.1016/j.corsci.2006.02.006.

Nwanonenyi, S., Obasi, M., Obidiegwu, H. & Chukwujike, I. 2020 Anticorrosion response of polymer mixture on mild steel in hydrochloric acid environment. *Emergent Materials* **3** (5), 663–673.

Olajire, A. A. 2013 *Fundamentals of Oilfield Chemistry for Professionals in Petroleum and Energy Industries*. Akmos Environmental Consult, Ogbomosho, Nigeria, p. 560.

Osorio-Celestino, G. R., Hernandez, M., Solis-Ibarra, D., Tehuacanero-Cuapa, S., Rodriguez-Gomez, A. & Gomora-Figueroa, P. 2020 Influence of calcium scalling on corrosion behavior of steel and aluminum alloys. *ACS Omega* **5**, 17304–17313. https://dx.doi.org/10.1021/acsomega.0c01538.

Ozgun, H., Ersahin, M. E., Erdem, S., Atay, B., Sayili, S., Eren, E., Hoshan, P., Atay, D., Altinbas, M., Kinaci, C. & Koyuncu, I. 2013 Comparative evaluation for characterization of produced water generated from oil, gas, and oil-gas production fields. *Clean-Soil, Air, Water* **41** (12), 1175–1182. https://doi.org/10.1002/clen.201200204.

Parangusan, H., Bhadra1, J. & Al-Thani, N. 2021 A review of passivity breakdown on metal surfaces: influence of chloride and sulfide-ion concentrations, temperature, and pH. *Emergent Materials* **4**, 1187–1203. https://doi.org/10.1007/s42247-021-00194-6.

Patil, D. B. & Sharma, A. R. 2011 Study on the corrosion kinetics of iron in acid and base medium. *E-Journal of Chemistry* **8** (1), 358–362. http://dx.doi.org/10.1155/2011/294792.

Piri, E. R., Shams, G., Shahmansouri, M. & Farzadkia, M. 2008 Survey of corrosion and scaling potential in drinking water of distribution system of Khoramabad city with corrosion indices. *Quarterly Newsletter* **10** (3). https://dx.doi.org/10.5681 %2Fhpp.2012.013.

Rajani, B., McDonald, S. & Felio, G. 1995 *Water Mains Break Data on Different Pipe Materials for 1992 and 1993. Report A-7019.1*. National Research Council of Canada, Ottawa, Canada.

Sawyer, C. L., McCarthy, P. L. & Parkin, G. F. 1994 *Chemistry for Environmental Engineering*, 4th edn. McGraw-Hill Book Company, New York, p. 545.
Short, K. C. & Stauble, A. J. 1967 Outline of geology of Niger Delta. *American Association of Petroleum Geologists Bulletin* **51**, 761–779.
Weber, K. J. & Daukoru, E. M. 1975 Petroleum geological aspects of the Niger Delta Nigeria. *Journal of Mineral Geology* **12**, 9–32.
Whiteman, A. 1982 *Nigeria: Its Petroleum Geology, Resources, and Potential*, Vols. 1 and 2. Graham and Trotman, London.

First received 10 August 2021; accepted in revised form 30 May 2022. Available online 3 June 2022

doi: 10.2166/wst.2022.067

Adsorption of heavy metals by *Lycium barbarum* branch-based adsorbents: raw, fungal modification, and biochar

Jian Guan[a], Changwei Hu[b], Jun Zhou[c], Qingguo Huang[d] and Jiayang Liu[a,*]

[a] College of Environmental Science and Engineering, Nanjing Tech University, Nanjing 211816, China
[b] College of Biological Chemical Science and Engineering, Jiaxing University, Jiaxing 314001, China
[c] College of Biotechnology and Pharmaceutical Engineering, Nanjing Tech University, Nanjing 211816, China
[d] Department of Crop and Soil Sciences, University of Georgia, Griffin, Georgia 30223, USA
*Corresponding author. E-mail: jyliu@njtech.edu.cn

ABSTRACT

The present study reports on the adsorptive removal of heavy metals (Cr^{3+}, Cd^{2+}, and Cu^{2+}) from water by a series of *Lycium barbarum* branch-based adsorbents: *Lycium barbarum* branch (denoted as LB), fungal fermented LB (FLB), LB biochar (LBB), FLB biochar (FLBB), alkaline modified LBB (ALBB), and alkaline modified FLBB (AFLBB). The six adsorbents were characterized in terms of FTIR, SEM, surface area and pore size as well as zeta potential. The adsorptive potential of these adsorbents was tested under varying conditions – pH, contact time, initial concentration, and temperature. Adsorption results were well fitted with Langmuir, Freundlich, and Temkin models. The maximum adsorption capacity (q_m) was calculated to be 6.29 mg/g for Cr^{3+} by FLB, 11.53 mg/g for Cd^{2+} by LB, and 7.27 mg/g for Cu^{2+} by LB. The pseudo-second-order kinetic equation better described the adsorption process. Based on the thermodynamics parameters, the adsorption of heavy metals was endothermic but not spontaneous for biochars. The experimental results offer a new way for recycling and reutilizing LB in wastewater treatment.

Key words: adsorption, biochar, fermentation, heavy metal, *Lycium barbarum* branch

HIGHLIGHTS

- *Lycium barbarum* branch (LB) was used to adsorb Cu^{2+}, Cr^{3+}, and Cd^{2+}.
- LB was further modified with fungal fermentation and O_2-free pyrolysis.
- Isotherm, kinetic, and thermodynamic of adsorption were fitted and analyzed.
- Modified LB-based adsorbents showed varied adsorption capacity and behavior.
- LB and FLB are novel and promising adsorbents.

GRAPHICAL ABSTRACT

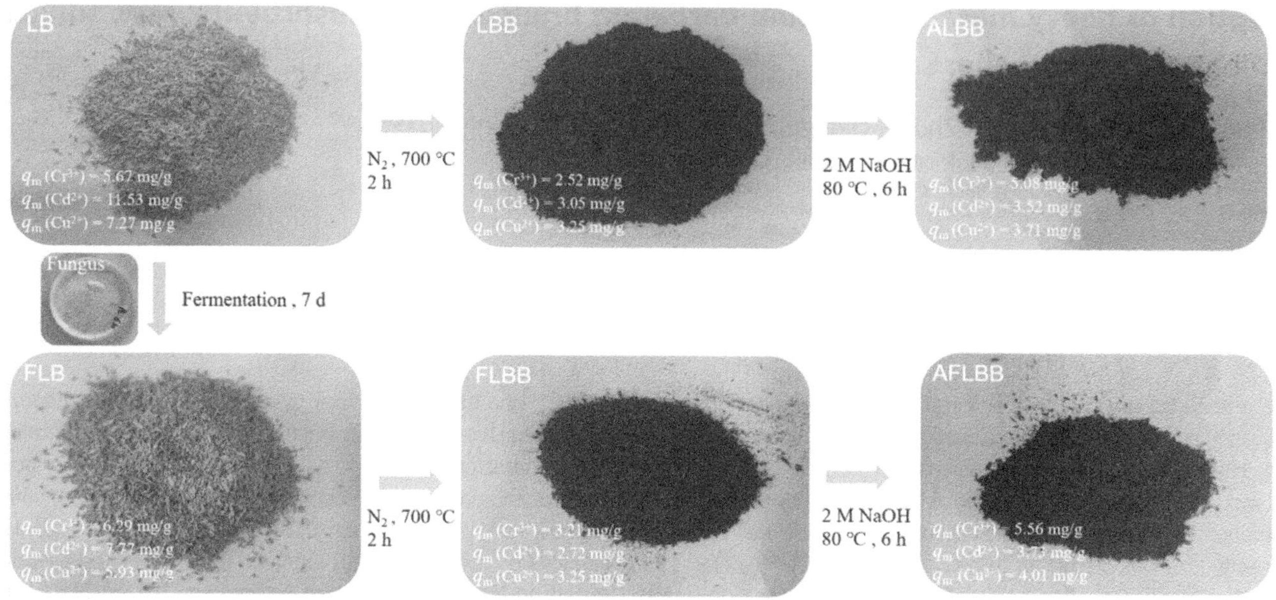

1. INTRODUCTION

Anthropogenic activities, i.e., leather, mining, textile, electroplating, battery, and pesticide processing/production as well as e-waste dumping are thought to be the dominant contributors for heavy metals in wastewater (Song & Li 2015). Although awareness in environmental protection has been increasingly reinforced worldwide, high concentrations of heavy metals can be still detected in natural waters. Luo *et al.* (2021) investigated a series of lakes along Yangtse River and revealed the heavy metals in sediments (mg/kg): Cr (90.8), Cu (60.1), Hg (0.06), Zn (102), Cd (0.89), Pb (42.7), and As (6.01). Heavy metals normally enter into water as ions, which not only jeopardize aquatic lives but also through food chains build up in human bodies, causing organ failure and even cancer (Balali-Mood *et al.* 2021).

Relatively little research has been conducted towards Cr^{3+} due to its low toxicity, which, however, could be enhanced by 100-fold when oxidized into Cr^{6+} in the environment, thus posing great threat to environmental ecology and human health (Jin *et al.* 2020). The presence of Cd^{2+} within the concentrations ranging from 0.18 to 50 µM could cause malformations, delay and arrest of development in a dose-dependent manner, with *P. wall* as a test subject (Calevro *et al.* 1998). Even though at low concentrations Cu^{2+} could foster the formation of red blood cells and maintaining nervous system and immune functions, an overdose, however, could do great harm to the human body by damaging kidney, brain, nervous and blood system (Vardhan *et al.* 2019).

To achieve efficient removal of heavy metals from water, many methods have been thoroughly examined, among which the adsorption technique stands out for its obvious advantages, e.g., flexible design, easy operation, and cost–effectiveness (Wong *et al.* 2018; Chai *et al.* 2021). Using lignocellulosic wastes as green adsorbents is gaining interest among researchers all over the world (Mo *et al.* 2018; Sahmoune 2019). Oiltea shell proves to possess the maximum adsorption capacity of 22.4 (Pb^{2+}), 12.1 (Cu^{2+}), and 14.2 mg/g (Cd^{2+}), respectively (Liu *et al.* 2019). Pretreatments of agro-wastes, e.g., modification of rice straw with *Comamonas testosteroni* FJ17 (Xue *et al.* 2020), can further improve the adsorptive capacity. In addition, activated carbons and biochars derived from agro-wastes have been widely tested as adsorbents as well (Wong *et al.* 2018; Deng *et al.* 2019).

Lycium barbarum (denoted as LB in this study) represents a typical Chinese herb and cash crop due to its pharmaceutical/food trait, whose planting area covers 82,000 hectares in northwest China (Masci *et al.* 2018; Alam *et al.* 2021). Periodic trimming and clipping of LB as well as berry harvesting co-produce a large quantity of branch waste, amounting to 0.2 million tons in a single year. The majority of these wastes are piled up in processing factories and basically burnt, which apparently badly affects air quality and the global zero-carbon emission goal. It is of great practical significance to reutilize LB biomass for its high availability. High content of cellulosic component in LB makes it a candidate as an adsorbent (Mo *et al.* 2018) or as feedstock to produce biochar/activated carbon (Danish & Ahmad 2018).

In this study, LB branches evaluated with respect to its adsorptive potential for Cr^{3+}, Cd^{2+}, and Cu^{2+} from water. To find out the most efficient adsorbent, LB branches treated in different ways, including fungal fermentation and pyrolysis as well as alkaline modification. To the best of our knowledge, this is the first study using LB branches as adsorbents.

2. MATERIALS AND METHODS

2.1. *Lycium barbarum* branches and reagents

Lycium barbarum branches (LB) were provided by a local farm in Ningxia Hui Autonomous Region, China. LB was air dried and grounded into powder for further use. Metal ions were made in solution using $CuCl_2 \cdot 2H_2O$, $CdCl_2 \cdot 5/2H_2O$, $Cr(NO_3)_3 \cdot 9H_2O$, respectively. These metal salts and other chemicals such as HNO_3 (65–68%), HCl, and NaOH were all purchased from the Sinopharm Chemical Reagent Co., Ltd (Shanghai, China).

2.2. Preparation of adsorbents

The powdered LB collected from the above was passed through an 80-mesh screen and oven dried at 80 °C to constant weight, then was directly used as an adsorbent or further treated to prepare modified adsorbents. Some of the LB were taken and mixed with water at a ratio of 1:4 and put in a glass Petri dish ($\Phi = 9$ cm) for sterilization at 121 °C and 0.12 MPa for 20 min. Afterwards, one piece of fungal dish of *Pleurotus ostretus* was cut and inoculated directly into the center of the autoclaved and cooled Petri dish (Liu *et al.* 2018). The culture lasted for 7 days at static conditions and 30 °C to allow complete spreading of fungal mycelia on the matrix. The fermented matrix was then soaked in deionized water for 4 h under shaking conditions, followed by separation of the solids using vacuum filtration. The separated solid was oven-dried at 80 °C as described above and labelled as FLB. LB and FLB were then subjected to pyrolysis for 2 h at 700 °C with an inlet gas of N_2. The biochars unloaded from the tube furnace were washed with 0.1 M HCl to clean the impurities on the surface and then further washed with deionized water to neutral. After oven drying at 80 °C, the prepared biochars were termed LBB and FLBB, respectively. LBB and FLBB were further modified with alkaline by immersing 6 g of them in 100 mL NaOH solution (2 mol/L) for 6 h in a 80 °C water bath (Liu *et al.* 2020). The alkaline modified biochars were separated from solution and washed with deionized water to neutral and oven dried at 80 °C, and the obtained solids were named ALBB and AFLBB, respectively. In total, six adsorbents were prepared in this study – LB, FLB, LBB, FLBB, ALBB, and AFLBB.

2.3. Adsorbent characterization

The chemical composition, i.e., cellulose, hemicellulose, lignin, and ash, of raw and fungal-treated LB, was analyzed using National Standards: GB/T 6434-2006, GB/T 6433-2006, GB/T6432-1994. The Brunauer–Emmett–Teller (BET) method was adopted to determine the specific surface area, total pore volume, and average pore size (Liu *et al.* 2018). Surface morphology of six different adsorbents was obtained using a scanning electronic microscopy (SEM). FTIR spectra of adsorbents were recorded within a range of 400–4,000 cm^{-1}. Zeta potential of adsorbents was obtained at pH = 6 with a Malvern Zetasizer Nano ZS90 instrument. COD and chromaticity of the solution where adsorption was carried out by different adsorbents (dosage 2 g/L, 12 h) were examined according to previous reports (He *et al.* 2018; Jing *et al.* 2018).

2.4. Effect of pH, temperature, and metal concentration on adsorption

To study the effect of pH value on adsorption, the metal solution was adjusted with 0.1 mol/L NaOH and/or HNO_3 within the range of pH 3–7. The adsorption was conducted in a system containing adsorbent 2 g/L and metal ion 10 mg/L in a 10 mL solution in a glass flask, which was stabilized on a rotatory shaker (160 rpm) at room temperature for 12 h. After the adsorption was finished, the solution was passed through a 0.45 μm filter and the concentration of the metal ions in the filtrate was determined by ICP-MS (iCAP Q, Thermo, USA). Three replicates of each adsorption experiment were conducted and the mean values were adopted for plotting with the standard deviation lower than 5%.

Through Equations (1) and (2), metal removal rate R (%) and amount of sorbed metal by adsorbent q (mg/g) were calculated:

$$R\ (\%) = \frac{C_0 - C_e}{C_0} \times 100 \tag{1}$$

$$q\ (mg/g) = \frac{(C_0 - C_e)V}{m} \tag{2}$$

where C_0 and C_e (mg/L) are the initial and equilibrium concentration of the ion, respectively; q (mg/g) is the amount of ion sorbed by adsorbent; V (L) is the initial volume of solution; and m (g) is the weight of the adsorbent.

2.5. Isotherms and dynamics of adsorption

In the reaction system stated above, initial metal concentration was adjusted to 2, 4, 6, 8, 10, 14, 18, 28, and 36 mg/L. The reaction was conducted at room temperature ($25 \pm 0.5\ °C$) for 12 h. Using the values calculated from Equations (1) and (2), Langmuir, Freundlich, and Temkin models were fitted using Equations (3), (4), and (5), respectively (Zhang *et al.* 2021):

$$\frac{C_e}{q_e} = \frac{1}{K_L q_m} + \frac{C_e}{q_m} \tag{3}$$

$$\lg q_e = \frac{1}{n}\lg c_e + \lg K_F \tag{4}$$

$$q_e = \frac{RT}{B_T \ln A_T} + \frac{RT}{BT \ln c_e} \tag{5}$$

where K_L (L/mg) is the Langmuir adsorption constant and q_m (mg/g) is the maximum ion amount of adsorption corresponding to complete monolayer coverage on the surface, q_e (mg/g) is the amount of ion adsorbed by sorbent at equilibrium, and C_e (mg/L) is the equilibrium concentration of ion solution. K_F is an indicator of adsorption capacity (mg/g) and 1/n is the adsorption intensity.

Using the following Equation (6) separation factor R_L was calculated, where C_0 was initial metal concentration (mg/L):

$$R_L = \frac{1}{(1 + KLco)} \tag{6}$$

In the reaction system and conditions stated above, the initial metal concentration was adjusted to 10 mg/L, and the reaction was conducted at room temperature ($25 \pm 0.5\ °C$) for varying durations −5, 10, 20, 40, 60, 120, 240, 360, and 720 min. After the adsorption, instant (q_t) and equilibrium adsorption capacity (q_e) were calculated as above stated. The data were then fitted with Lagergren's pseudo-first-order model (7), Ho's pseudo-second-order model (8), and the intraparticle diffusion model (9) (Liu *et al.* 2018):

$$\lg (q_e - q_t) = \lg q_e - \frac{K_1 t}{2.303} \tag{7}$$

$$\frac{t}{q_t} = \frac{1}{K_2 q_e^2} + \frac{t}{q_e} \tag{8}$$

$$q_t = K_3 \sqrt{t} + c \tag{9}$$

where q_e is equilibrium adsorption amount (mg/g); q_t is adsorption amount (mg/g) at contact time t (h or min); K_1 is the equilibrium rate constant of the first order sorption (min^{-1}); K_2 is the equilibrium rate constant of the second order sorption (g/mg·min). K_3 is the intraparticle rate constant (mg/g·min$^{1/2}$); and C is the film diffusion extent (mg/g).

2.6. Thermodynamic analysis

Adsorbent at dosage of 2 g/L was added into a glass flask containing 10 mL metal solution (pH = 6, 10 mg/L) to start the reaction. The adsorption was carried out at 298 K, 308 K, and 318 K on a rotator at 160 rpm for 12 h. Afterwards, the solution was passed through a 0.45 μm filter for metal quantification, from which final metal concentration (C_e) and adsorption capacity (q_e) were calculated. The change in free energy (ΔG^o) was calculated using the following equation to study the thermodynamic nature:

$$\Delta G^o = -RT \ln(q_e/C_e) \tag{10}$$

where R is the gas constant (8.3143 J/mol K), and T is the absolute temperature, q_e is equilibrium adsorption amount (mg/g); C_e is equilibrium ion concentration (mg/L).

From the plot of ΔG^o vs. T, the value of enthalpy ΔH^o and entropy ΔS^o can be calculated as follows (Liu & Lee 2014):

$$\Delta G^o = \Delta H^o - T\Delta S^o \tag{11}$$

---

3. RESULTS AND DISCUSSION

3.1. Adsorbent characterization

As shown in Table 1, the contents of cellulose, hemicellulose, and lignin in raw LB were 25.5, 14.3, and 31.5%, respectively, which was reduced to 21.1, 13.8, and 29.6%, respectively, in fungal-treated LB (i.e., FLB). This indicates that LB, as an agro-waste, has great potential in serving as a novel matrix for fungal growth. Once fungal growth occurred, fungal fermentation-based modification of biosorbents would become possible due to changes in chemical composition and, as a result, the adsorptive capacity might be enhanced or behavior altered (Liu *et al.* 2016; Liu *et al.* 2018). Fungal modification has another advantage – it produces value-added co-product, namely ligninolytic enzymes that find broad application in wastewater treatment (Wang *et al.* 2019).

Physical properties of the six adsorbents were characterized and the results are shown in Table 2. Drastic variation in specific surface area (SSA) was observed among adsorbents. Raw LB held the smallest SSA of 0.086 m^2/g, which was elevated to 0.109 m^2/g for fungal-treated LB (i.e., FLB). Pyrolysis of LB and FLB led to a steep increase in SSA of LBB and FLBB, reaching 18.963 m^2/g and 19.958 m^2/g, respectively. Alkaline treatment of biochars caused a further increase in SSA of ALBB and AFLBB, reaching 27.174 m^2/g and 28.225 m^2/g, respectively, suggesting the feasibility of using alkaline to modify carbon-based materials (Zhao *et al.* 2020). An increasing trend was also observed for total pore volume (cm^3/g) with the order of LB < FLB < LBB < FLBB < ALBB < AFLBB. An opposite trend was found with average pore size (APS), which was decreased along with the various treatments. Carbonization via pyrolysis downsized APS (~350 nm) of LB and FLB to approximately 20 and 35 nm of biochars. Zeta potentials for LB, FLB, LBB, FLBB, ALBB, and AFLBB at pH 6.0 were measured as: −18.40, −17.57, −24.07, −25.80, −24.80, and −26.70 mV, respectively, which were in good accordance with other reports (Kong *et al.* 2014; Liu *et al.* 2018; Zhang *et al.* 2021).

Functional groups on the surface of the six adsorbents were revealed by FTIR spectra in Figure 1. LB as well as FLB was found to be rich in typical absorbance peaks on the surface, which was reduced significantly after pyrolysis for preparation of biochars. The phenomenon of attenuation of numerous functional groups in biochars has been reported (Zhao *et al.* 2021). Stretching the band of O-H at wavenumber 3,600–3,150 cm^{-1} in LB and FLB indicated the vibration of hydroxyl groups present in cellulose, hemicellulose, and lignin (Hashem *et al.* 2020). Dehydration at high temperature during pyrolysis would lead to a reduction in hydroxyl groups in biochars. A small peak at 2,930 cm^{-1} and 2,860 cm^{-1} corresponding to C-H was observed in LB and FLB (Zhang *et al.* 2020). A peak at 1,740 cm^{-1} corresponded to the ester C = O structure, while the

Table 1 | Chemical composition of *Lycium barbarum* branch (LB) and fermented *Lycium barbarum* branch (FLB)

	Component (%)			
Material	**Cellulose**	**Hemicellulose**	**Lignin**	**Ash**
LB	25.5	14.3	31.5	6.5
FLB	21.1	13.8	29.6	14.0

Table 2 | Specific surface area, pore volume, pore size, and zeta potential of six different adsorbents

Adsorbent	Specific surface area (m²/g)	Pore volume (cm³/g)	Median pore width (nm)	Zeta potential (mV)
LB	0.086	0.001	357.275	−18.40
FLB	0.109	0.002	340.807	−17.57
LBB	18.963	0.014	29.824	−24.07
FLBB	19.958	0.015	19.859	−25.80
ALBB	27.174	0.040	28.288	−24.80
AFLBB	28.225	0.035	49.535	−26.70

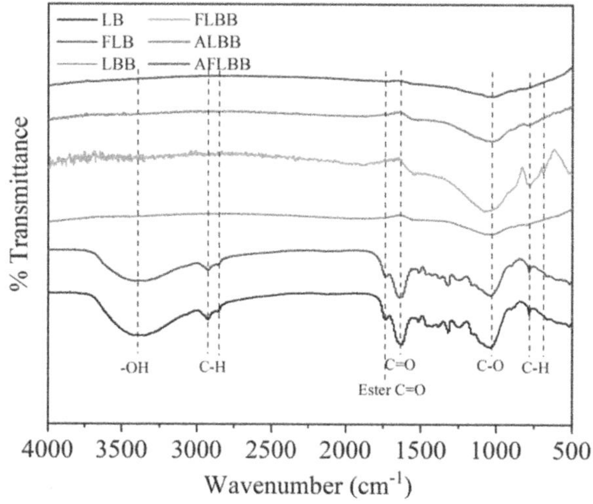

Figure 1 | FTIR spectra of six different adsorbents.

peak at 1,620 cm^{-1} corelated with the phenolic C=O functional group (Xu *et al.* 2020). The strong stretching vibration of C-O at 1,040 cm^{-1} could be caused by hemicellulose and lignin (Gan *et al.* 2016), which accounted for a higher intensity in LB and FLB than that in four biochars. Moreover, the weak bending vibrations of C-H at 780 cm^{-1} and 690 cm^{-1} in LB and FLB were enhanced in FLBB.

Figure 2 portrays the SEM images of adsorbents at magnification of ×1,000. From the micro-structures in these graphs, it was relatively smooth on the surface of the raw LB, with few ravines and cavities being seen (Figure 2(a)). Fungal treatment clearly damaged the smoothness of the raw LB surface and increased the heterogeneous morphology (Ighalo & Adeniyi 2020), i.e., more ridged ravines and cavities (Figure 2(b)). As for biochars derived from biomasses, very tiny flaky particles, which looked like impurities and were inferred to mean the possible presence of cavities by some authors (Ighalo & Adeniyi 2020), appeared on the surface of LBB and FLBB (Figure 2(c) and 2(d)). In addition, more ruptured surfaces with cracks and crevices on biochars have been observed. With further treatment with alkaline, these particles on biochars were partly removed, which made the surface look much cleaner, presumably exposing more micro-structures for the increase in SSA and pore volume of ALBB and AFLBB (Figure 2(e) and 2(f) and Table 2).

3.2. Effect of several parameters on adsorption

Profiles of metal adsorption by the six adsorbents at varying conditions are revealed in Figure 3. pH value could significantly impact adsorption by adjusting the degree of ionization of metals as well the distribution of surface charge on adsorbent. Within the pH range 3–7, a general trend that higher pH value led to higher metal removal rate was observed among six adsorbents (Wang *et al.* 2015). Comparatively, an order of removal efficiency can be seen: LB/FLB > ALBB/AFLBB > LBB/FLBB. At pH 3, abundant H$^+$ in the solution would compete with metal ions on adsorption sites, thus resulting in a low removal rate. When the pH of the solution was increased to 7, a steep increase in metal removal took place, largely due to the formation of precipitable hydroxides (Zhao *et al.* 2020). Conversely, the charge on adsorbents became more

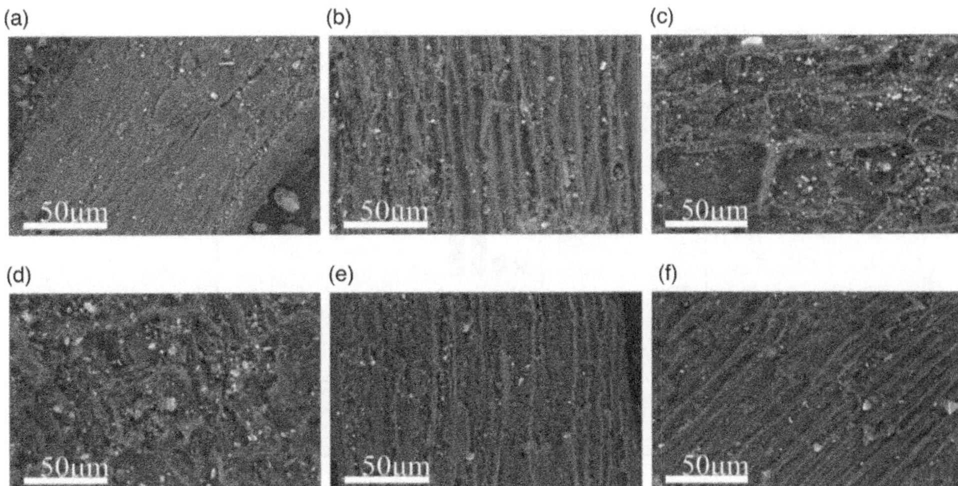

Figure 2 | SEM images of six different adsorbents: (a) LB, (b) FLB, (c) LBB, (d) FLBB, (e) ALBB, (f) AFLBB.

negative when the pH was increased from 3 to 6, facilitating the electrostatic attraction between metal cations and the negatively charged surface (Kong *et al.* 2014). This is especially significant for LB and FLB (Figure 3) because they had more functional groups (i.e., -COOH) than the biochars (Figure 1). The reaction solution was therefore adjusted to pH = 6 in the following experiments.

As demonstrated in Figure 4, contact time notably influenced the metal removal rate. Similar to the results in most of the literature, over half of the adsorption was finished in the beginning of the process, e.g., 10 min for the adsorption of Cr^{3+} and Cd^{2+} by LB and FLB, indicating rapid occupation of available adsorptive sites. Along with reaction proceeded, equilibrium was gradually achieved due to the decreased number of adsorptive sites. Adsorption reached equilibrium after 120 min, and this was therefore chosen as the contact time in the other adsorption studies. Removal rate by biochars (LBB and FLBB) was apparently lower than that by LB and FLB when the reaction was just initiated, possibly due to aggregation of biochars in solution. After a short while, the removal rate returned to normal, probably because metals were re-adsorbed onto the adsorption sites following a transient desorption.

Initial metal ion concentration also affected the adsorption by adsorbents in a severe way (Figure 5). There was a general trend among six adsorbents as metal removal rate would constantly decrease as initial metal concentration was increased gradually. For instance, at initial concentration of 2 mg/L, the removal rates for Cr^{3+} by LB, FLB, LBB, FLBB, ALBB, and AFLBB were 85.59, 91.42, 41.23, 50.23, 62.93, and 77.29%, respectively, which was reduced to 24.75, 27.24, 11.04, 13.17, 19.48, and 24.16%, correspondingly, when the initial concentration was increased to 36 mg/L. A similar trend was found for Cd^{2+}, and Cu^{2+}. By contrast, with increasing initial metal ion concentration, adsorption capacity (q_e) at equilibrium showed an increasing momentum (see inserted figures in Figure 5). With respect to removal of Cr^{3+}, Cd^{2+}, and Cu^{2+}, LB and FLB performed better than alkaline modified biochars (ALBB and AFLBB), which were better than biochars (LBB and FLBB).

3.3. Isotherms of adsorption

Langmuir, Freundlich, and Temkin models were applied in this study and the corresponding parameters were calculated as shown in Table 3. For adsorption of Cr^{3+}, the Langmuir model yielded a higher R^2 than the Freundlich model, suggesting the monolayer adsorption of Cr^{3+} on six adsorbents. Similar findings have been reported using biomass pine as adsorbent (Zhao *et al.* 2021). As for Cd^{2+}, in contrast, the Freundlich model fitted well with the adsorption compared with the Langmuir model, implying the occurrence of adsorption onto the heterogeneous surfaces (Xu *et al.* 2015). Both Langmuir and Freundlich models could well describe the adsorption of Cu^{2+} on the six adsorbents. The separation factor K_L in the Langmuir model was less than 1, indicating that the adsorption of Cr^{3+}, and Cd^{2+}, and Cu^{2+} was favorable ($0 < K_L < 1$). In the Freundlich model, values of $1/n$ for the three ions were all <1, suggesting that the adsorption of heavy metals by six adsorbents was

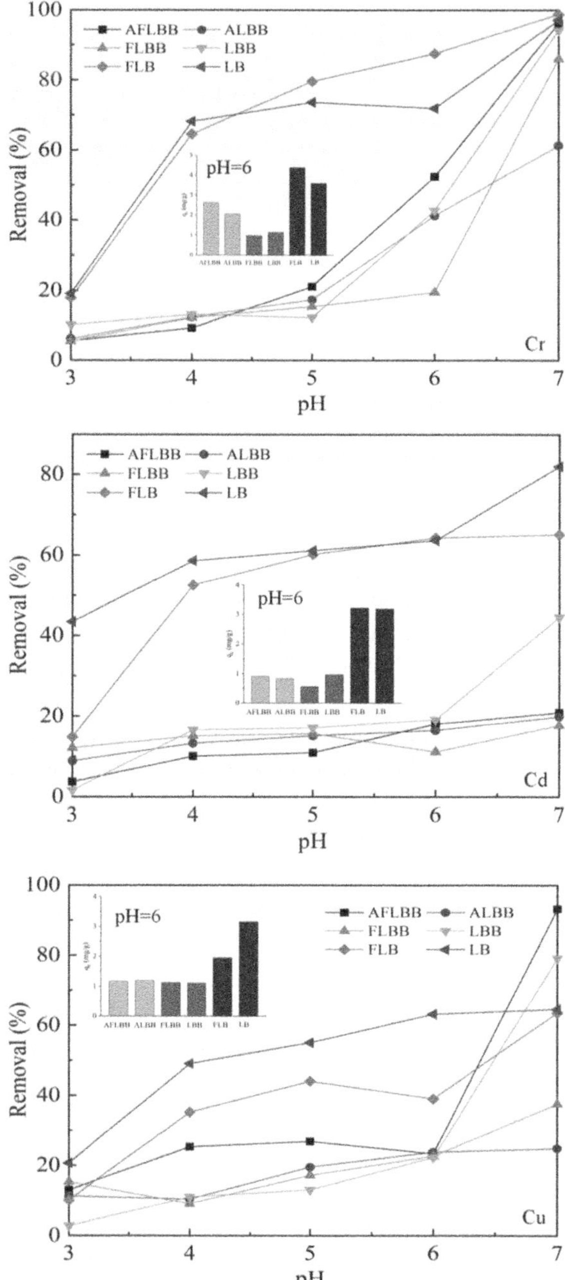

Figure 3 | Effect of pH value of solution on heavy metal removal (metal 10 mg/L, sorbent 2 g/L, 25 °C, 160 rpm, 12 h). Each inserted figure is the equilibrated adsorption capacity (mg/g) following 12 h contact at pH = 6.

very feasible. Further, Temkin models for six adsorbents generated a high R^2 of 0.92, indicating the existence of strong intermolecular forces in the process of adsorption (Zhang *et al.* 2021).

Generally, raw LB and fungal-fermented LB (i.e., FLB) showed better adsorption performance towards three heavy metals. Pyrolysis of LB into biochars did not enhance adsorption, while further alkaline modification of biochars seemed to improve slightly. According to the Langmuir model, the maximum adsorption capacity (q_{m}) was calculated to be 6.29 mg/g for Cr^{3+} by FLB, 11.53 mg/g for Cd^{2+} by LB, and 7.27 mg/g for Cu^{2+} by LB. Comparison of maximum adsorption capacity of heavy metals by different agro-adsorbents in previous reports and in this study was made (Table 4). For the raw LB, its adsorption capacity fell well in the range of previously reported results among various agro-wastes. As for fungal pretreatment of LB (i.e., FLB), it was apparently better than some of raw agro-/industrial wastes, e.g., poplar and pine, and comparable with some

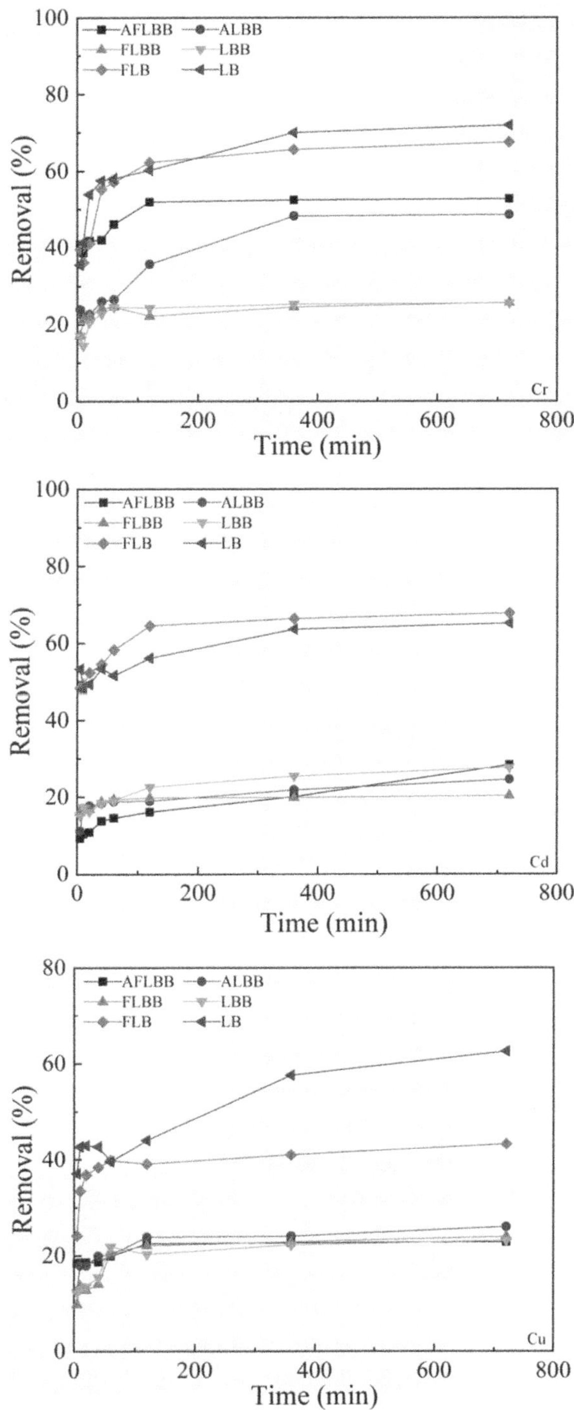

Figure 4 | Effect of contact time on heavy metal removal (pH = 6, sorbent 2 g/L, 25 °C, 160 rpm, metal 10 mg/L).

modified ones, such as modified peanut hush. This reveals that LB and FLB have great potential as adsorbents to remove heavy metals from water. From the perspective of removal efficiency and energy input, more research should be carried out with biochars and even alkaline modified biochars.

3.4. Dynamics of adsorption

Three kinetic models, namely pseudo-first-order (PFO), pseudo-second-order (PSO), and intraparticle diffusion (IPD), were applied to fit the adsorption process and the related parameters are listed in Table 5. By comparing the values

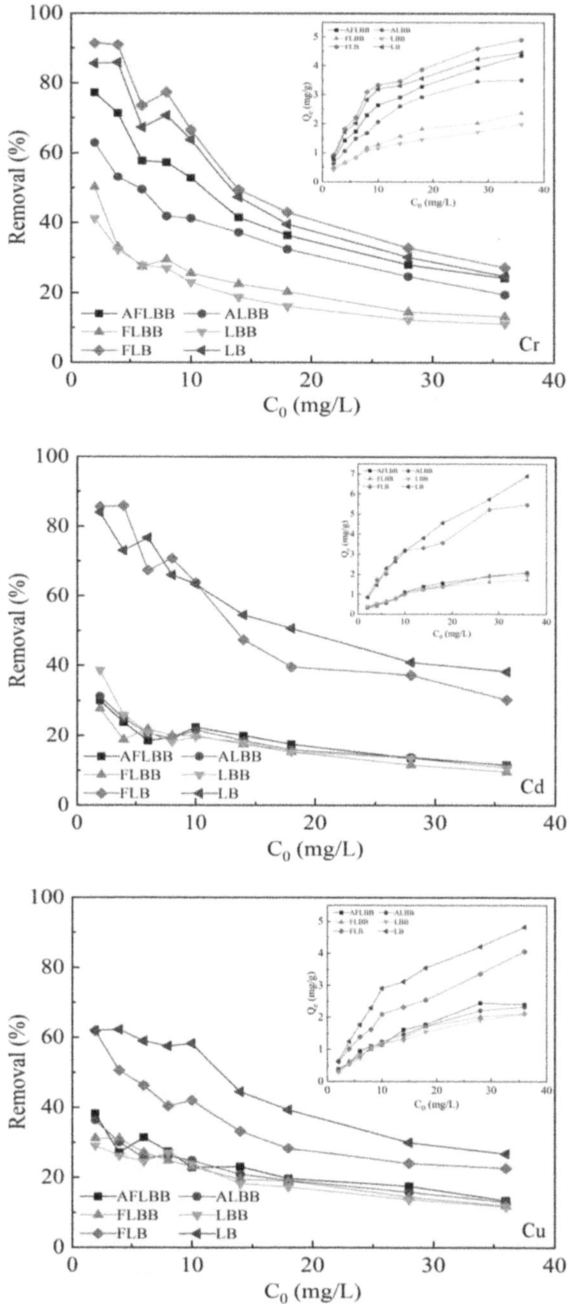

Figure 5 | Effect of initial concentration on heavy metal removal (pH = 6, sorbent 2 g/L, 25 °C, 160 rpm, 12 h). Each inserted figure is the equilibrated adsorption capacity (mg/g) following 12 h contact at different initial metal concentrations.

of R^2 for all the adsorption experiments, PSO ($R^2 > 0.958$) better explained the adsorption process, suggesting that the adsorption of these heavy metals on six adsorbents belonged to chemical adsorption. Furthermore, the calculated capacity ($q_{e\ cal}$) based on PSO model was closer to the experimental results ($q_{e\ exp}$). Comparatively, raw LB showed high R^2 values for PFO models but still lower than those for PSO model, which was not found in the other five LB-based adsorbents. Many adsorption processes using agro-wastes can be well described with both PFO and PSO models (Lee & Choi 2018). Using cow manure-derived biochars as adsorbents for Cu^{2+} removal, adsorption processes can be well fitted with both PFO and PSO models (Zhang *et al.* 2021), which is slightly different from our results of LB-based biochars. Diffusion processes of metals transferred from aqueous solution onto adsorbents can be described

Table 3 | Parameters of Langmuir, Freundlich, and Temkin adsorption models for six different adsorbents

Adsorbent	Langmuir			Freundlich			Temkin		
	K_L (L/mg)	q_m (mg/g)	R^2	K_F (mg/g)	n	R^2	A_T (L/g)	B_T (kJ/mol)	R^2
LB-Cr	0.1033	5.67	0.9893	0.7567	1.85	0.9180	1.00	1.96	0.9826
FLB-Cr	0.0968	6.29	0.9893	0.7998	1.82	0.9254	0.96	1.79	0.9859
LBB-Cr	0.0855	2.52	0.9875	0.3159	1.89	0.9790	0.89	4.60	0.9822
FLBB-Cr	0.0692	3.21	0.9714	0.3317	1.77	0.9781	0.75	3.67	0.9605
ALBB-Cr	0.0689	5.08	0.9898	0.4638	1.63	0.9725	0.71	2.26	0.9773
AFLBB-Cr	0.0827	5.56	0.9962	0.6202	1.76	0.9609	0.83	2.04	0.9909
LB-Cd	0.0377	11.53	0.9803	0.5677	1.40	0.9874	0.54	1.19	0.9525
FLB-Cd	0.0622	7.77	0.9516	0.6753	1.64	0.9585	0.71	1.55	0.9542
LBB-Cd	0.0496	3.05	0.9181	0.2313	1.64	0.9801	0.61	4.16	0.9227
FLBB-Cd	0.0519	2.72	0.9560	0.1812	1.48	0.9600	0.62	4.49	0.9735
ALBB-Cd	0.0404	3.52	0.9480	0.1952	1.46	0.9869	0.55	3.80	0.9437
AFLBB-Cd	0.0372	3.73	0.8997	0.1813	1.39	0.9710	0.53	3.65	0.9462
LB-Cu	0.0540	7.27	0.9749	0.4816	1.46	0.9520	0.64	1.68	0.9847
FLB-Cu	0.0502	5.93	0.9593	0.4264	1.57	0.9892	0.62	2.13	0.9468
LBB-Cu	0.0582	3.25	0.9764	0.2192	1.51	0.9622	0.62	3.85	0.9762
FLBB-Cu	0.0556	3.25	0.9922	0.2363	1.53	0.9740	0.64	3.73	0.9770
ALBB-Cu	0.0487	3.71	0.9855	0.2442	1.51	0.9875	0.60	3.39	0.9637
AFLBB-Cu	0.0460	4.01	0.9442	0.2472	1.48	0.9748	0.58	3.17	0.9481

by the IPD model. Some of the adsorptions could be well described by the IPD model based on R^2 values, such as Cr^{3+} by AFLBB, Cd^{2+} by FLB, LB, and ALBB, and Cu^{2+} by FLB, FLBB, and AFLBB, while the others had very poor fitting. Taking adsorption of Cu^{2+} by AFLBB as an example, such processes might be separated into three phases (Zhang *et al.* 2021): diffusion phase ($K_{d1} = 0.03$), adsorption phase ($K_{d2} = 0.014$), and equilibrium ($K_{d3} = 0.001$), thus further verifying the saturation of free electric potential on adsorbents.

3.5. Thermodynamics of adsorption

In the profile of temperature effect on adsorption (figure not shown), elevated removal rate and adsorption capacity were observed at higher temperatures. The reason for this phenomenon might be ascribed to increased odds of collision between metal ions and adsorbents as well as more exposed adsorption sites on adsorbent surface at higher temperature. The nature of endothermic or exothermic inclination, spontaneity, randomness, and the temperature favorability for sorption process can be identified using thermodynamic analysis (Mahmoud *et al.* 2021). Gibbs free energy ($\Delta G°$) was calculated based on Equation (10) and plotted vs. temperature, from which enthalpy change $\Delta H°$ and entropy change $\Delta S°$ were obtained according to Equation (11) and these values are summarized in Table 6. The negative values of $\Delta G°$ support the favorability of the adsorption process. With a few exceptions (adsorption with LB and FLB), most adsorptions in this study generated positive $\Delta G°$ values, meaning that the adsorption processes by these adsorbents were not spontaneous. Adsorption of heavy metals ions by green adsorbents has been reported to be spontaneous in most cases ($\Delta G° < 0$) (Sahmoune 2019). In addition, a decreasing trend was found for $\Delta G°$ along with an increasing temperature in all adsorption scenarios, implying that extra heat in solution might favor the adsorption. $\Delta H°$ and $\Delta S°$ values for all the adsorption processes were found to be positive, in the range of 6.9 to 44.9 kJ·mol^{-1} and 0.013 to 0.157 kJ·mol^{-1}·K^{-1}, respectively. Adsorption of Cd^{2+} by the *Cynara scolymus*-derived biochar showed values of $\Delta G°$ (-6.28 kJ/mol), $\Delta H°$ (13.47 kJ/mol), and $\Delta S°$ (59.6 kJ·mol^{-1}·K^{-1}), respectively (Mahmoud *et al.* 2021). The positive values of $\Delta G°$ together with positive values of $\Delta H°$ strongly indicated that the sorption of heavy metals by these adsorbents was nonspontaneous and endothermic. The positive values of $\Delta S°$ suggest varied affinity of

Table 4 | Comparison of maximum adsorption capacity of heavy metals by different adsorbents in previous reports and in this study

Metal ion	Adsorbent	$q_{max\ (mg/g)}$ (*T* (°C)/pH)	References
Cr^{3+}	Modified peanut husk	7.67 (25/4)	Li *et al.* (2007)
	Poplar	5.52 (25/4)	Wang *et al.* (2015)
	Pine	1.14 (25/4)	Zhao *et al.* (2021)
	FLB	6.29 (25/6)	This study
Cd^{2+}	Persimmon leaf	22.59 (25 /6)	Lee & Choi (2018)
	Coffee husk	6.90 (25/4)	Oliveira *et al.* (2008)
	Linden	3.50 (room temperature/5.1)	Božić *et al.* (2009)
	LB	11.53 (25/6)	This study
Cu^{2+}	Modified peanut husk	10.15 (25/4)	Li *et al.* (2007)
	Poplar	6.59 (25/4)	Li *et al.* (2007)
	Persimmon leaf	19.42 (25/6)	Lee & Choi (2018)
	LB	7.27 (25/6)	This study

Table 5 | Parameters of pseudo-first-order, pseudo-second-order, and intraparticle diffusion kinetic adsorption models by six different adsorbents

Adsorbent	PFO				PSO			IPD		
	$q_{e\ exp}$ (mg/g)	K_1 (min^{-1})	$q_{e\ cal}$ (mg/g)	R^2	K_2 (g/mg min^{-1})	$q_{e\ cal}$ (mg/g)	R^2	K_3 (mg/g h$^{-1/2}$)	C (mg/g)	R^2
LB-Cr	3.60	0.0074	1.31	0.9194	0.0238	3.64	0.9993	0.0369	2.67	0.7458
FLB-Cr	3.37	0.0075	1.12	0.8325	0.0294	3.41	0.9998	0.0166	2.94	0.9555
LBB-Cr	1.28	0.0090	0.27	0.7005	0.1510	1.29	0.9999	0.0041	1.18	0.8174
FLBB-Cr	1.29	0.0034	0.18	0.1686	0.1318	1.28	0.9990	0.0111	1.00	0.9294
ALBB-Cr	2.43	0.0124	1.79	0.9640	0.0137	2.53	0.9945	0.0406	1.44	0.5493
AFLBB-Cr	2.64	0.0113	0.58	0.8157	0.0694	2.66	0.9998	0.0024	2.57	0.9607
LB-Cd	3.26	0.0064	0.84	0.9483	0.0323	3.28	0.9992	0.0286	2.54	0.751
FLB-Cd	3.39	0.0075	0.82	0.8669	0.0437	3.41	0.9998	0.0103	3.11	0.9881
LBB-Cd	1.38	0.0047	0.57	0.9500	0.0397	1.40	0.9974	0.0162	0.95	0.9934
FLBB-Cd	1.02	0.0053	0.13	0.5408	0.2694	1.02	0.9998	0.0021	0.96	0.7464
ALBB-Cd	1.23	0.0033	0.42	0.7128	0.0508	1.23	0.9954	0.0177	0.75	0.9994
AFLBB-Cd	1.42	0.0021	0.86	0.8939	0.0149	1.40	0.9582	0.0383	0.35	0.9140
LB-Cu	3.13	0.0041	1.23	0.9047	0.0149	3.16	0.9940	0.0584	1.63	0.8728
FLB-Cu	2.16	0.0040	0.41	0.4349	0.0706	2.16	0.9992	0.0132	1.80	0.9913
LBB-Cu	1.15	0.0066	0.38	0.5566	0.0709	1.17	0.9992	0.0091	0.92	0.9116
FLBB-Cu	1.20	0.0070	0.48	0.6927	0.0556	1.21	0.9989	0.0058	1.04	0.9773
ALBB-Cu	1.30	0.0047	0.41	0.6281	0.0666	1.30	0.9985	0.0068	1.10	0.6378
AFLBB-Cu	1.14	0.0098	0.21	0.7568	0.1798	1.15	0.9998	0.0010	1.40	0.9908

heavy metals toward different adsorbents and the increased randomness at the solid–solution interface (Lim *et al.* 2016).

3.6. Adsorption mechanism

The order of adsorption capacity has been determined as: LB&FLB > ALBB&AFLBB > LBB&FLBB, which is supposed to be the result of combined effects of several factors – SSA, pore volume, pore size, zeta potential, and functional groups.

Table 6 | Parameters of thermodynamic of adsorption by six different adsorbents

Adsorbent T (K)	Thermodynamic parameters				
	ΔG° (kJ mol^{-1})			ΔH° (kJ mol^{-1})	ΔS° (kJ mol^{-1} K^{-1})
	298	308	318		
LB-Cr	−2.318	−2.650	−5.168	44.927	0.157
FLB-Cr	−1.867	−2.409	−2.874	16.025	0.060
LBB-Cr	3.003	2.785	2.566	6.949	0.013
FLBB-Cr	3.154	2.352	2.175	15.663	0.042
ALBB-Cr	2.595	2.396	2.118	7.561	0.017
AFLBB-Cr	−0.277	−0.441	−0.861	9.234	0.032
LB-Cd	−1.321	−1.750	−2.378	16.766	0.061
FLB-Cd	−1.786	−2.617	−2.982	19.084	0.070
LBB-Cd	3.416	2.885	2.405	16.075	0.042
FLBB-Cd	3.738	2.922	2.522	19.397	0.053
ALBB-Cd	3.193	2.730	2.362	13.222	0.034
AFLBB-Cd	2.880	2.692	2.344	8.530	0.019
LB-Cu	−1.197	−1.836	−2.194	15.889	0.057
FLB-Cu	0.794	0.408	0.234	8.941	0.027
LBB-Cu	3.268	2.704	2.171	17.434	0.048
FLBB-Cu	3.192	2.833	2.239	15.112	0.040
ALBB-Cu	2.747	2.487	2.071	10.724	0.027
AFLBB-Cu	3.016	2.616	2.255	12.110	0.031

Although LB and FLB have the lowest SSA ($\sim$0.1 m^2/g), their average pore width is the largest ($\sim$350 nm). In addition, the most abundant functional groups are present on the surface of LB and FLB, which could increase the possibility of chelating with heavy metal ions. AFLBB, holding the biggest SSA (28.225 m^2/g) and pore volume (0.035 cm^3/g) and highest negative zeta potential (-26.7 mV) but the lest functional groups, eventually exhibited median adsorption capacity among six adsorbents. LBB and FLBB have less functional groups and median SSA and pore size and finally exhibit the poorest adsorption capability. It follows that functional groups, such as -COOH and -OH, might play a dominant role in adsorption of heavy metal ions from water which is accompanied with other minor factors, such as zeta potential and pore size.

3.7. COD and chromaticity of the adsorption solution

Direct use of agro-wastes as adsorbents may be accompanied with organic and nutrient-leaching problems, thus increasing the contamination risk to aquatic systems (Kong *et al.* 2014). COD and chromaticity of the adsorption solution using six adsorbents are shown in Figure 6. With dosage of 2 g/L, LB (raw) generated COD and chromaticity of 99 mg/L and 420, respectively, while FLB (with fungal modification) significantly reduced the values to 43 mg/L and 93, respectively, showing the advantage of FLB as an adsorbent for metal removal with less introduction of organics from adsorbent. Comparatively, all the biochars as well as alkaline modified ones did not exhibit such problems because no COD and chromaticity were detected in the adsorption solution. For practical application, adsorption efficiency, energy input, potential leaching contamination, etc., should be taken into consideration.

4. CONCLUSIONS

High availability of LB together with the experimental results make it possible to be utilized as a novel adsorbent in its raw or modified form for removing heavy metals from water. By comparing the adsorption capacity, LB and FLB (fungal treatment) outcompeted the others (biochars) even though the latter seem to have enhanced specific surface area and pore volume that resulted from pyrolysis. It can be deduced that active sites might govern the adsorption of heavy metal ions in a greater

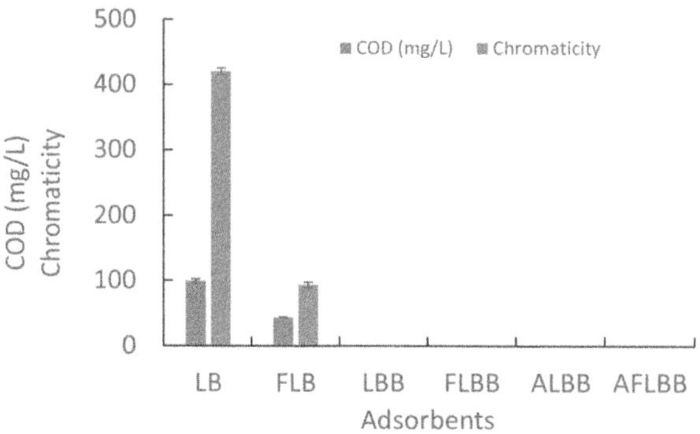

Figure 6 | COD and chromaticity of adsorption solution with different adsorbents (sorbent 2 g/L, 12 h).

manner because LB and FLB apparently possess more functional groups on the surface than do the corresponding biochars. Through the potential contamination leaching test, FLB proved be more advantageous relative to LB because the former released less COD and chromaticity into the aqueous system. In addition, it is economically feasible to manufacture FLB as an adsorbent as the fungal treatment normally co-produces value-adding enzymes.

ETHICAL APPROVAL

Not applicable.

CONSENT TO PARTICIPATE

Not applicable.

CONSENT TO PUBLISH

Not applicable.

AUTHORS CONTRIBUTIONS

JG carried out the experiments, analyzed the data, and drafted the manuscript. CH helped analyze the metal concentration using ICP-MS. JZ provided the adsorbent materials and proposed the idea. QH and JL designed and supervised the experiments as well as edited/revised the manuscript. All authors read and approved the final manuscript.

FUNDING

This work was supported by the National Natural Science Foundation of China (No. 51503074), the Key Research and Development Technology of Ningxia Hui Autonomous Region (special project for foreign science and technology cooperation, 2019BFH02008), the Jiangsu Agriculture Science and Technology Innovation Fund (SCX(21)2175), Graduate Student Scientific Research Innovation Projects in Jiangsu Province (SJCX21_0463), and the Key Research and Development Technology of Shanxi (201903D211013).

COMPETING INTERESTS

The authors declare that they have no competing interests.

DATA AVAILABILITY STATEMENT

All relevant data are included in the paper or its Supplementary Information.

REFERENCES

Alam, M. A., Muhammad, G., Khan, M. N., Mofijur, M., Lv, Y., Xiong, W. & Xu, J. 2021 Choline chloride-based deep eutectic solvents as Green extractants for the isolation of phenolic compounds from biomass. *J. Clean. Prod.* **309**, 127445.

Balali-Mood, M., Naseri, K., Tahergorabi, Z., Khazdair, M. R. & Sadeghi, M. 2021 Toxic mechanisms of five heavy metals: mercury, lead, chromium, cadmium, and arsenic. *Front. Pharmacol.* **12**, 643972.

Božić, D., Stanković, V., Gorgievski, M., Bogdanović, G. & Kovačević, R. 2009 Adsorption of heavy metal ions by sawdust of deciduous trees. *J. Hazard. Mater.* **171**, 684–692.

Calevro, F., Campani, S., Ragghianti, M., Bucci, S. & Mancino, G. 1998 Tests of toxicity and teratogenicity in biphasic vertebrates treated with heavy metals (Cr^{3+}, Al^{3+}, Cd^{2+}). *Chemosphere* **37**, 3011–3017.

Chai, W. S., Cheun, J. Y., Kumar, P. S., Mubashir, M., Majeed, Z., Banat, F., Ho, S.-H. & Show, P. L. 2021 A review on conventional and novel materials towards heavy metal adsorption in wastewater treatment application. *J. Clean. Prod.* **296**, 126589.

Danish, M. & Ahmad, T. 2018 A review on utilization of wood biomass as a sustainable precursor for activated carbon production and application. *Renew. Sust. Energ. Rev.* **87**, 1–21.

Deng, Y., Huang, S., Laird, D. A., Wang, X. & Meng, Z. 2019 Adsorption behaviour and mechanisms of cadmium and nickel on rice straw biochars in single- and binary-metal systems. *Chemosphere* **218**, 308–318.

Gan, W., Gao, L., Zhan, X. & Li, J. 2016 Preparation of thiol-functionalized magnetic sawdust composites as an adsorbent to remove heavy metal ions. *RSC Adv.* **6**, 37600–37609.

Hashem, A., Badawy, S. M., Farag, S., Mohamed, L. A., Fletcher, A. J. & Taha, G. M. 2020 Non-linear adsorption characteristics of modified pine wood sawdust optimised for adsorption of Cd(II) from aqueous systems. *J. Environ. Chem. Eng.* **8**, 103966.

He, Y., Song, N. & Jiang, H.-L. 2018 Effects of dissolved organic matter leaching from macrophyte litter on black water events in shallow lakes. *Environ. Sci. Pollut. Res.* **25**, 9928–9939.

Ighalo, J. O. & Adeniyi, A. G. 2020 A mini-review of the morphological properties of biosorbents derived from plant leaves. *SN Appl. Sci.* **2**, 509.

Jin, M., Xiao, X., Qin, L., Geng, W., Gao, Y., Li, L. & Xue, J. 2020 Physiological and morphological responses and tolerance mechanisms of *isochrysis galbana* to Cr(VI) stress. *Bioresour. Technol.* **302**, 122860.

Jing, L., Chen, B., Wen, D., Zheng, J. & Zhang, B. 2018 The removal of COD and NH_3-N from atrazine production wastewater treatment using UV/O_3: experimental investigation and kinetic modeling. *Environ. Sci. Pollut. Res.* **25**, 2691–2701.

Kong, Z., Li, X., Tian, J., Yang, J. & Sun, S. 2014 Comparative study on the adsorption capacity of raw and modified litchi pericarp for removing Cu(II) from solutions. *J. Environ. Manage.* **134**, 109–116.

Lee, S.-Y. & Choi, H.-J. 2018 Persimmon leaf bio-waste for adsorptive removal of heavy metals from aqueous solution. *J. Environ. Manage.* **209**, 382–392.

Li, Q., Zhai, J., Zhang, W., Wang, M. & Zhou, J. 2007 Kinetic studies of adsorption of Pb(II), Cr(III) and Cu(II) from aqueous solution by sawdust and modified peanut husk. *J. Hazard. Mater.* **141**, 163–167.

Lim, L. B. L., Priyantha, N., Chieng, H. I. & Dahri, M. K. 2016 *Artocarpus camansi* Blanco (Breadnut) core as low-cost adsorbent for the removal of methylene blue: equilibrium, thermodynamics, and kinetics studies. *Desalin. Water Treat.* **57**, 5673–5685.

Liu, X. & Lee, D.-J. 2014 Thermodynamic parameters for adsorption equilibrium of heavy metals and dyes from wastewaters. *Bioresour. Technol.* **160**, 24–31.

Liu, J., Li, E., You, X., Hu, C. & Huang, Q. 2016 Adsorption of methylene blue on an agro-waste oiltea shell with and without fungal treatment. *Sci. Rep.* **6**, 38450.

Liu, J., Wang, Z., Li, H., Hu, C., Raymer, P. & Huang, Q. 2018 Effect of solid state fermentation of peanut shell on its dye adsorption performance. *Bioresour. Technol.* **249**, 307–314.

Liu, J., Hu, C. & Huang, Q. 2019 Adsorption of Cu^{2+}, Pb^{2+}, and Cd^{2+} onto oiltea shell from water. *Bioresour. Technol.* **271**, 487–491.

Liu, J., Yang, X., Liu, H., Cheng, W. & Bao, Y. 2020 Modification of calcium-rich biochar by loading Si/Mn binary oxide after NaOH activation and its adsorption mechanisms for removal of Cu(II) from aqueous solution. *Colloids Surf. A* **601**, 124960.

Luo, M., Yu, H., Liu, Q., Lan, W., Ye, Q., Niu, Y. & Niu, Y. 2021 Effect of river-lake connectivity on heavy metal diffusion and source identification of heavy metals in the middle and lower reaches of the Yangtze River. *J. Hazard. Mater.* **416**, 125818.

Mahmoud, M. E., Abou-Ali, S. A. A. & Elweshahy, S. M. T. 2021 Efficient and ultrafast removal of Cd(II) and Sm(III) from water by leaves of *Cynara scolymus* derived biochar. *Mater. Res. Bull.* **141**, 111334.

Masci, A., Carradori, S., Casadei, M. A., Paolicelli, P., Petralito, S., Ragno, R. & Cesa, S. 2018 *Lycium barbarum* polysaccharides: extraction, purification, structural characterisation and evidence about hypoglycaemic and hypolipidaemic effects. A review. *Food Chem.* **254**, 377–389.

Mo, J., Yang, Q., Zhang, N., Zhang, W., Zheng, Y. & Zhang, Z. 2018 A review on agro-industrial waste (AIW) derived adsorbents for water and wastewater treatment. *J. Environ. Manage.* **227**, 395–405.

Oliveira, W. E., Franca, A. S., Oliveira, L. S. & Rocha, S. D. 2008 Untreated coffee husks as biosorbents for the removal of heavy metals from aqueous solutions. *J. Hazard. Mater.* **152**, 1073–1081.

Sahmoune, M. N. 2019 Evaluation of thermodynamic parameters for adsorption of heavy metals by green adsorbents. *Environ. Chem. Lett.* **17**, 697–704.

Song, Q. & Li, J. 2015 A review on human health consequences of metals exposure to e-waste in China. *Environ. Pollut.* **196**, 450–461.

Vardhan, K. H., Kumar, P. S. & Panda, R. C. 2019 A review on heavy metal pollution, toxicity and remedial measures: current trends and future perspectives. *J. Mol. Liq.* **290**, 111197.

Wang, H., Gao, B., Wang, S., Fang, J., Xue, Y. & Yang, K. 2015 Removal of Pb(II), Cu(II), and Cd(II) from aqueous solutions by biochar derived from KMnO4 treated hickory wood. *Bioresour. Technol.* **197**, 356–362.

Wang, F., Xu, L., Zhao, L., Ding, Z., Ma, H. & Terry, N. 2019 Fungal laccase production from lignocellulosic agricultural wastes by solid-state fermentation: a review. *Microorganisms* **7**, 665.

Wong, S., Ngadi, N., Inuwa, I. M. & Hassan, O. 2018 Recent advances in applications of activated carbon from biowaste for wastewater treatment: a short review. *J. Clean. Prod.* **175**, 361–375.

Xu, X., Bai, B., Wang, H. & Suo, Y. 2015 Enhanced adsorptive removal of Safranine T from aqueous solutions by waste sea buckthorn branch powder modified with dopamine: kinetics, equilibrium, and thermodynamics. *J. Phys. Chem. Solids* **87**, 23–31.

Xu, Z., Xu, X., Zhang, Y., Yu, Y. & Cao, X. 2020 Pyrolysis-temperature depended electron donating and mediating mechanisms of biochar for Cr(VI) reduction. *J. Hazard. Mater.* **388**, 121794.

Xue, C., Zhang, Q., Owens, G. & Chen, Z. 2020 A cellulose degrading bacterial strain used to modify rice straw can enhance Cu(II) removal from aqueous solution. *Chemosphere* **256**, 127142.

Zhang, Y., Jiao, X., Liu, N., Lv, J. & Yang, Y. 2020 Enhanced removal of aqueous Cr(VI) by a green synthesized nanoscale zero-valent iron supported on oak wood biochar. *Chemosphere* **245**, 125542.

Zhang, P., Zhang, X., Yuan, X., Xie, R. & Han, L. 2021 Characteristics, adsorption behaviors, Cu(II) adsorption mechanisms by cow manure biochar derived at various pyrolysis temperatures. *Bioresour. Technol.* **331**, 125013.

Zhao, N., Li, B., Huang, H., Lv, X., Zhang, M. & Cao, L. 2020 Modification of kelp and sludge biochar by TMT-102 and NaOH for cadmium adsorption. *J. Taiwan Inst. Chem. Eng.* **116**, 101–111.

Zhao, J., Boada, R., Cibin, G. & Palet, C. 2021 Enhancement of selective adsorption of Cr species via modification of pine biomass. *Sci. Total Environ.* **756**, 143816.

First received 4 November 2021; accepted in revised form 11 February 2022. Available online 24 February 2022

doi: 10.2166/wst.2022.250

Regeneration of As(V) loaded granular activated carbon through desorption in $FeCl_3$, $CaCl_2$ and $MgCl_2$ aqueous solutions

Niels Michiel Moed * and Young Ku

Department of Chemical Engineering, National Taiwan University of Science and Technology, Taipei, Taiwan (R.O.C.)
*Corresponding author. E-mail: nikolai.moed@gmail.com

NMM, 0000-0002-4448-3425

ABSTRACT

As(V) adsorption on granular activated carbon (GAC) and subsequent desorption in dH_2O was modeled using the pseudo-first and pseudo-second order kinetic models. Regeneration was achieved by immersing loaded GAC in NaCl, $FeCl_3$, $CaCl_2$ and $MgCl_2$ aqueous solutions. As(V) detection after desorption was highest for NaCl but subsequent adsorption was lowest. Regeneration was highest in $FeCl_3$ solution of pH 2 followed closely by pH 3, but As(V) precipitation appeared superior at pH 3. Molar ratios of Fe, Ca and Mg to As were tested in the range of 0.75:1 to 12:1 where a logarithmic relation was found between the molar ratio and As(V) desorption as diluted in HNO_3 and H_2O and subsequent adsorption. Precipitation was nearly complete in $FeCl_3$, limited in $MgCl_2$ at a ratio of 12:1 and not observed in $CaCl_2$. While kinetic values were lower than in previous tests, the pseudo-first and pseudo-second order models could accurately describe desorption in $CaCl_2$ and $MgCl_2$ but not in $FeCl_3$ due to precipitation. Desorption in $FeCl_3$ was most effective in precipitating As(V), being highest at a molar ratio of 6:1, but regeneration was slightly higher at a molar ratio of 12:1.

Key words: activated carbon, arsenic, precipitation, regeneration, wastewater

HIGHLIGHTS

- Regeneration of activated carbon through desorption of As(V) was combined with coprecipitation.
- 94.7% regeneration was reached using ferric chloride solution.
- Coprecipitation with Fe(III) was able to remove As(V) from the solution down to 0.2 mg/L.
- Desorption of As(V) from activated carbon in water was kinetically modeled.
- Desorption in media other than water ($FeCl_3$, $CaCl_2$ and $MgCl_2$) was kinetically modeled for the first time.

GRAPHICAL ABSTRACT

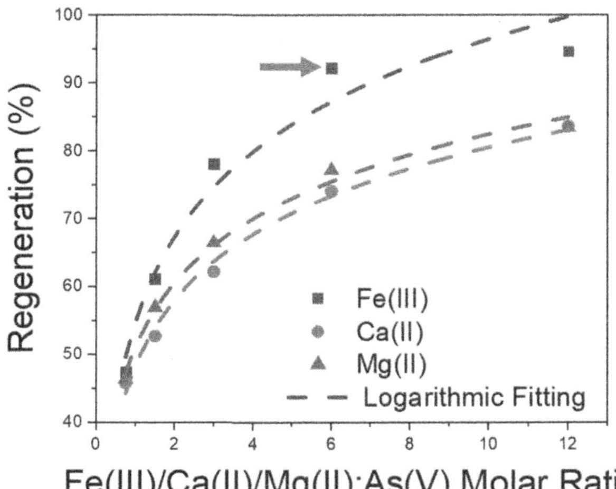

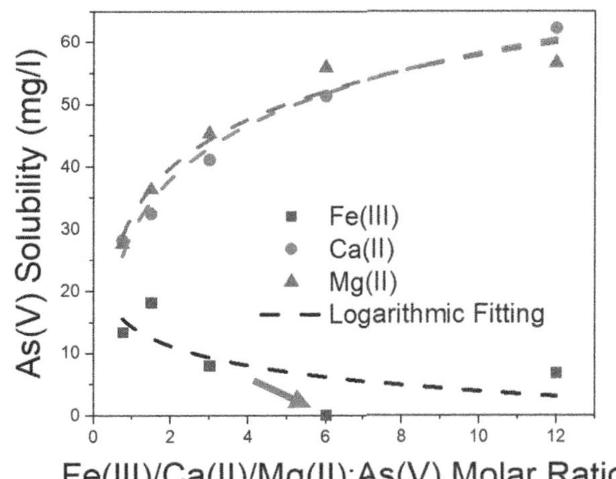

ABBREVIATIONS

GAC Granular activated carbon
PFO Pseudo-first order
PSO Pseudo-second order

1. INTRODUCTION

Arsenic is widely distributed throughout the Earth's environment (Mandal & Suzuki 2002). Exposure to arsenic causes a wide array of health problems including respiratory (Milton *et al.* 2001), hematological (Armstrong *et al.* 1984) and carcinogenic (Banerjee *et al.* 2011). Many papers report on removal of arsenic from water and soil using a variety of methods such as filtration (Waypa *et al.* 1997), electrocoagulation (Nidheesh & Anantha Singh 2017), precipitation and adsorption (Gupta & Chen 1978).

Adsorption is the adhesion of an adsorbate onto the surface of an adsorbent. Common adsorbents for arsenic adsorption are iron hydroxide (Pierce & Moore 1982), aluminium hydroxide (Anderson *et al.* 1976) and activated carbon (Lorenzen *et al.* 1995), but other materials can be used, such as diatomite (Davoodi *et al.* 2019) and walnut shell (Ashrafi *et al.* 2018). The surface of activated carbon is typically porous, leading to an extraordinarily large specific surface area and pore volume. The properties of the pores and the functional groups on the surface depend on the preparation of the AC and determine its affinity to certain adsorbents. Over time the adsorption goes down as the adsorbent surface becomes saturated, at which point it is disposed of or regenerated (San Miguel *et al.* 2001; Ania *et al.* 2005). This regeneration can take place thermally, which is straightforward but expensive to operate and causes loss of adsorbent (Martin & Ng 1985; Salvador & Jiménez 1996; Hong *et al.* 2020). Nonthermal methods include the use of solvents, surfactants and acids or base. This last method is based on a change in pH, with the final pH being unfavorable to adsorption (McLaughlin 1995). One literature review indicated that research on desorption of cadmium and lead typically focus on regeneration through use of acids such as nitric acid, EDTA and hydrochloric acid (Fouda-Mbanga *et al.* 2021). Chemical regeneration is a versatile option, where desorption is improved through a change in chemical polarity, hydophobicity, solubility, molecular weight, pH, boiling point or toxicity (Larasati *et al.* 2021). It is important to consider the environmental fate of the desorbed contaminant, ensuring continued removal from waste streams. This means that an adsorption-desorption cycle should involve either a concentration of the contaminant or its separation. One way that separation can be achieved is through precipitation.

Precipitation is a common way to remove arsenic from aqueous waste streams. This is commonly achieved with iron (Halder *et al.* 2018), aluminium (Pantuzzo *et al.* 2014), calcium (Bothe & Brown 1999a; Palfy *et al.* 1999) and magnesium (Park *et al.* 2010). The solubility of precipitates is highly dependent on the solution pH, the species of both arsenic and the metal used for precipitation. In previous work arsenate removal from aqueous solution was tested through coagulation with $FeCl_2$, $FeCl_3$, $AlCl_3$, $CaCl_2$ and $MgCl_2$ (Moed & Ku 2022). For desorption and precipitation to simultaneously occur, it is important that favorable pH for these processes overlap. In this research, desorption was found to be optimal at an initial pH of 11, followed by a pH of 3. This excludes the use of $FeCl_2$, with optimal precipitation taking place at pH 8.5 (Johnston & Singer 2007) and $AlCl_3$, at pH 6 (Majzlan *et al.* 2018). Reported optimum precipitation of $FeCl_3$ varies from paper to paper but is generally shown to occur in the range of pH 2–5 (Nishimura & Tozawa 1978; Robins 1987). Precipitation with $CaCl_2$ increases with pH (Zhang *et al.* 2015), being optimal at pH over 12 (Bothe & Brown 1999a), whereas precipitation with Mg(II) was found to be optimal between pH 7.5 and 10.2 (Park *et al.* 2010).

The scope of this research is to investigate the use of metals to precipitate As(V) as it is desorbed through a shift in pH. The precipitation reduces the arsenic in the solution, which may lead to an increased desorption. The arsenic is simultaneously removed from the solution. Kinetic study will be performed for both adsorption and desorption of As(V) by and from the activated carbon used.

2. METHODS

2.1. Materials

Granular activated carbon (GAC) of 8×30 mesh (First Chemical Works in Taiwan) was washed with 0.3% HCl solution for 2 h at a GAC concentration of 100 g/L. It was then rinsed repeatedly with dH_2O, filtered through Advantec Grade No. 2 filter paper and dried overnight at 110 °C.

Na$_2$HAsO$_4$·7H$_2$O (Panreac, Spain), NaCl (Fischer, UK), FeCl$_3$·6H$_2$O (Acros Organics, USA), CaCl$_2$·2H$_2$O (Showa Chemical, Japan) and anhydrous MgCl$_2$ (Showa Chemical, Japan) were dissolved in dH$_2$O to create separate stock solutions. HCl and NaOH solutions at varying concentrations were used for pH adjustment. Experiments were, unless otherwise mentioned, performed inside a thermostatic shaking bath (DK-60 by Deng Yng, Taiwan) set to a shaking speed of 150 rpm and a temperature of 25 °C. All modeling was performed nonlinearly using the application Origin Pro.

The surface characteristics were tested by Brunauer, Emmett and Teller method (BET) (BELSORP-max by BEL Japan), Field Emission Scanning Electron Microscopy with Energy Dispersive Spectrometry (FE-SEM with EDS) (JSM-7900F by JEOL) and X-ray Fluorescence (XRF) (Epsilon 1 by Malvern Panalytical).

2.2. Adsorption of As(V)

Adsorption of As(V) on the fresh GAC used in this study was previously tested and found to be optimal in a solution with pH 6 (not reported). Different concentrations (50, 100, 150 and 200 mg/L) of As(V) were tested at a GAC concentration of 20 g/L, with samples being taken at set times to be filtered and analyzed by ICP-OES (iCAP 7000 by Thermo Scientific). The adsorption isotherm was tested at different concentrations of As(V) at pH 6 in plastic tubes of 50 mL volume.

2.3. Desorption in dH$_2$O

GAC was loaded with As(V) aqueous solution at a concentration of 200 mg/L and at pH 6. After a 20-h adsorption the GAC was filtered and dried overnight at 110 °C. Desorption of As(V) from the loaded GAC was tested in dH$_2$O at varying initial pH, where the pH was adjusted with HCl and NaOH solution. Desorption took place induplicate in 250 mL glass laboratory flasks. Samples were taken at set intervals, filtered and analyzed by ICP-OES.

2.4. Influence of pH

The influence of pH was tested on the regeneration of As(V) loaded GAC. The selected desorption solutions were FeCl$_3$, CaCl$_2$ and MgCl$_2$ at a molar concentration of Fe(III)/Ca(II)/Mg(II) equal to 1.5 times that of the As(V) loaded onto the GAC. The concentration of As(V) is calculated as the adsorption capacity after adsorption (4.00 mg/g) multiplied by the concentration of GAC in the regeneration (20 g/l) for a total of 80 mg/L. Desorption was tested in FeCl$_2$ at initial pH (pHi) of 2.0, 2.5 and 3.0 and in CaCl$_2$ and MgCl$_2$, separately, at pHi 11.0, 11.5 and 12.0. NaCl was added as a comparison at pHi 2.0, 3.0, 11.0 and 12.0. Desorption took place overnight after which the GAC was filtered and dried at 110 °C. Subsequent adsorption in 200 mg/L As(V) solution of pH 6 was performed overnight to determine regeneration efficiency. The desorption was performed in triplicate, to ensure accuracy, whereas subsequent adsorption was performed in duplicate, in either case using 50 mL plastic tubes. Samples of adsorption and desorption were taken which were diluted in HNO$_3$, filtered and measured by ICP-OES. The pH of each sample was measured after both desorption and adsorption, to test the influence of the GAC surface acidity on the desorption and adsorption media.

2.5. Influence of molar ratio

The influence of molar ratio was tested on the regeneration of As(V) loaded GAC was performed by adding GAC at a concentration of 20 g/L in aqueous solution of either Fe(III), Ca(II) or Mg(II) at molar ratios of 0.75:1, 1.5:1, 3:1, 6:1 and 12:1 of Fe(III)/Ca(II)/Mg(II):As(V). NaCl was found to be least effective in the previous test and therefore omitted from further tests. Due to a combination of factors (the As(V) desorbed, regeneration efficiency and perceived As(V) precipitation) the optimum pH$_i$ were chosen to be 3 for regeneration in FeCl$_3$ and 11 in.CaCl$_2$ and MgCl$_2$. Desorption was performed over 4 h, as this was found to be sufficient to reach maximum desorption, in duplicate in 50 mL plastic tubes. Two types of samples were taken, with the first diluted in HNO$_3$ and filtered to test the total As(V) desorption. The other type of samples was filtered prior to dilution in H$_2$O to test As(V) precipitation. The GAC was filtered and dried at 110 °C to test subsequent adsorption, and both desorption and subsequent adsorption were performed in 50 mL plastic tubes.

Kinetic desorption studies of As(V) from loaded GAC were performed at the highest ratios of 3:1, 6:1 and 12:1 of metal to arsenic. These kinetic tests were performed in 250 mL glass laboratory flasks at 25 °C, with samples taken at timed intervals. The data was modeled using the PFO and PSO kinetic models.

2.6. Error analysis

To provide a measure of certainty for the kinetic and isotherm analyses, the sum of squared residuals (SSR) was calculated as demonstrated (Batool *et al.* 2018; Herbert & Kumar 2021):

$$SSR = \sum_{i=1}^{n} (qe, i, calc - \ qe, i, meas)^2 \tag{1}$$

3. RESULTS AND DISCUSSION

3.1. Adsorption

Figure 1 reveals that it takes over 1 day for adsorption to be complete. The adsorption capacities for initial As(V) concentrations of 50, 100, 150 and 200 mg/L lead to adsorption capacities of 2.13, 3.16, 3.72 and 4.00 mg/g AC, respectively. The low difference in the equilibrium adsorption between initial concentrations of 150 and 200 mg/L As(V) suggests that maximum adsorption is nearly reached. Kinetics of the adsorption were modeled according to the pseudo-first order (PFO) and the pseudo-second order (PSO) adsorption kinetics. The PFO, as proposed by Lagergren, is commonly expressed in its nonlinear form as:

$$q_t = q_e(1 - e^{-k_1 t}) \tag{2}$$

where q_t is the adsorption capacity at a given time, generally in mg of adsorbate per g of adsorbent, q_e the adsorption capacity at equilibrium, k_1 the PFO rate constant and t is time (Lagergren 1898).

The PSO, popularized by the analyses of Ho and McKay, in its nonlinear form, can be written as:

$$q_t = \frac{q_e^2 k_2 t}{q_e k_2 t + 1} \tag{3}$$

where k_2 is the PSO kinetic rate constant (Ho & McKay 1999).

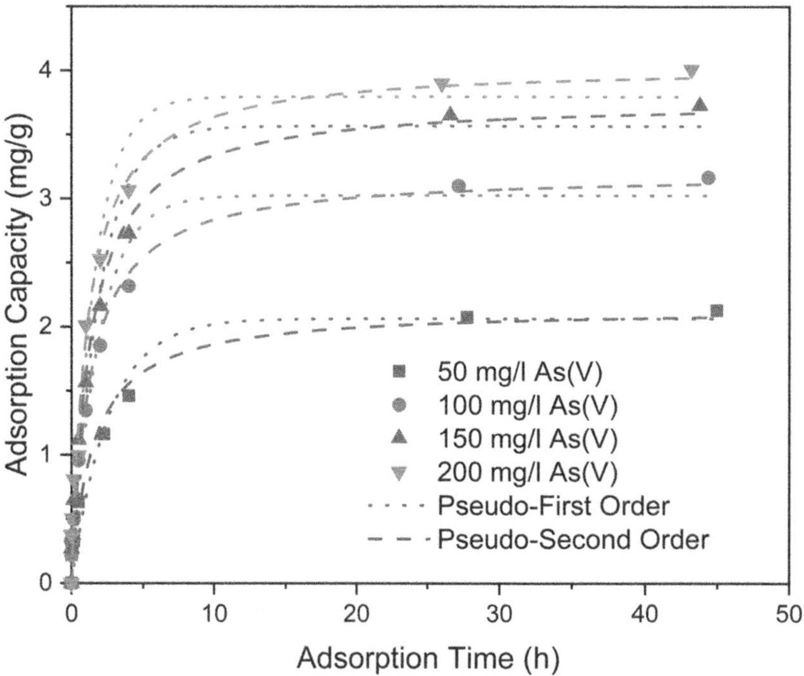

Figure 1 | Kinetic fitting for adsorption of As(V) concentrations on GAC at 20 g/L GAC, pH 6, 25 °C and a stirring speed of 150 rpm.

The results suggest that the PSO is a more accurate fit, as confirmed in Table 1 which reveals that, at any initial As(V) concentration, the PSO is most accurate in describing the adsorption of As(V) on GAC. This is revealed from the sum of the squared residuals which are, in all cases, lower for the PSO model than for the PFO model. This is also backed by the higher adjusted R^2 (over 0.96 for PSO and 0.94 for PFO) and from the comparison of the equilibrium adsorption capacity q_e found experimentally to those determined by the models. In their publication of 1999, Ho and McKay concluded that for all data analyzed and previously described using other kinetic models, the PSO was more accurate. The following popularity of the model has only increased over time as the PSO has been favored in the majority of kinetic modeling performed since (Ho 2006; Sun *et al.* 2013; Hassan *et al.* 2014; Koohzad *et al.* 2019). It appears that this superiority is so widely accepted that some papers do not even report on the PFO kinetic model (Yao *et al.* 2014; Gong *et al.* 2015).

The adsorption capacity (Q) for GAC has been plotted against initial As(V) concentration, as revealed in Figure 2. The Freundlich, Langmuir and Temkin adsorption isotherms were tested for GAC at various initial As(V) concentrations. The Freundlich isotherm model is described, in its nonlinear form, as:

$$q_e = K_F C_e^{1/n} \tag{4}$$

where K_F is the Freundlich constant, n the adsorption intensity and C_e (in mg/L) the equilibrium concentration of adsorbate (Al-Ghouti & Da'ana 2020).

Table 1 | Adsorption kinetics at 20 g/L GAC, pH 6, 25 °C and a stirring speed of 150 rpm

		Pseudo-first order				Pseudo-second order			
	qe_{exp}	qe	k1	R^2	RSS	qe	k2	R^2	SSR
50 mg/L	2.12623	2.0624	0.38532	0.94453	0.28227	2.1434	0.30906	0.96934	0.15600
100 mg/L	3.16429	3.02294	0.52232	0.96309	0.44857	3.20262	0.23932	0.98716	0.15604
150 mg/L	3.72720	3.56646	0.51289	0.96263	0.62941	3.77819	0.19939	0.98650	0.22743
200 mg/L	4.00812	3.79399	0.61990	0.95944	0.80015	4.04453	0.22059	0.98092	0.37648

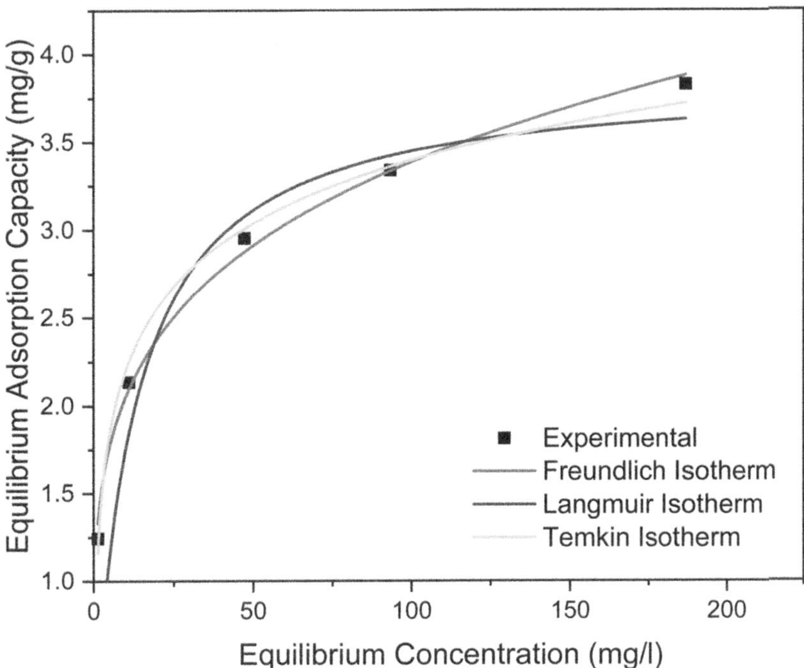

Figure 2 | Adsorption Isotherm modeling at 20 g/L GAC, pH 6, 25 °C and a stirring speed of 150 rpm.

The Langmuir isotherm model can be described nonlinearly as:

$$q_e = \frac{q_{max} K_L C_e}{1 + K_L C_e} \tag{5}$$

where q_{max} is the maximum adsorption capacity at complete monolayer coverage and K_L is the Langmuir constant in L/g (Langmuir 1916).

The Temkin isotherm model is expressed nonlinearly as:

$$q_e = B_T ln K_T + B_T ln C_e \tag{6}$$

where the constant B_T, which is related to the heat adsorption, is defined by the expression $B_T = RT/b$. In this expression, T the temperature in Kelvin, R the gas constant of 8.314 (J/mol K) and b the equilibrium constant related to the adsorption energy (in J/mol). K_T is the Temkin isotherm constant in L/g (Johnson & Arnold 1995). The closest fit seems to be the Freundlich isotherm, followed by the Temkin isotherm. Table 2 confirms this with the lower SSR values and the higher adjusted R^2 of the Freundlich model being 0.99455, the Temkin model 0.98878 and the Langmuir model 0.87506.

3.2. Desorption in dH$_2$O

Figure 3 reveals that desorption is highest at 43.8% in aqueous solution of pH$_i$ 11, followed by pH$_i$ 3. This is unsurprising, considering that adsorption was optimal at a pH$_i$ of 6. In the range of pH$_i$ 3–9, adsorption is near complete after 120 min.

Table 2 | Adsorption isotherm data at 20 g/L GAC, pH 6, 25 °C and a stirring speed of 150 rpm

Freundlich				Langmuir				Temkin			
kF	n	R²	SSR	kL	qm	R²	SSR	kT	B$_T$	R²	SSR
1.24047	4.58804	0.99455	0.01708	0.08239	3.86594	0.87506	38.30473	6.81068	4761.045	0.98878	0.03514

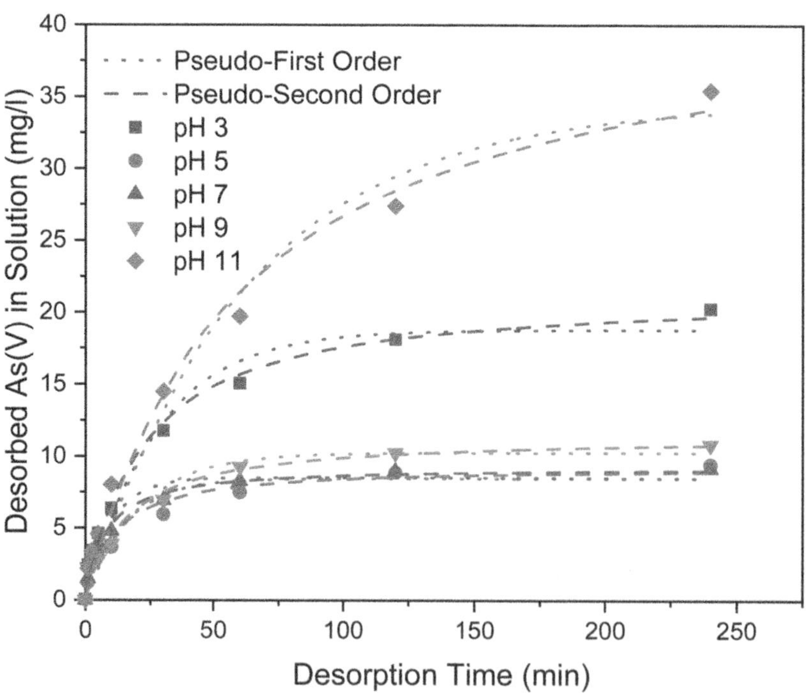

Figure 3 | Kinetic fitting for desorption of As(V) from loaded GAC in H$_2$O of varying pH at 20 g/L GAC loaded with 4.00 mg/g As(V), 25 °C and a stirring speed of 150 rpm.

At pH_i 11 this takes considerably longer, with equilibrium adsorption likely only measured once. Nonlinear PFO and PSO modeling was performed, where the adsorption capacity (q) of the previously used model was replaced by the As(V) concentration (C). When regenerating loaded ACCA (activated carbon-chitosan complex adsorbent) in a mixture of 1 mol/L thiourea and 2 mol/L hydrochloric acid, Ge & Fan (2011) found that 85.6% of lead and 88.2% of cadmium was desorbed. It was calculated, however, that the concentration of hydrochloric acid used was enough to lower the solution pH to below 1.8. In comparison, the solution pH within this test was closer to the pH that is favorable to adsorption. The results reveal that the PSO generally was a better fit except for desorption at pH_i 11, which is likely due to desorption being incomplete. Table 3 confirms that the PSO is more accurate in describing the desorption of As(V) from loaded GAC at every pH_i, with an R^2 of over 0.96 for all but pH_i 5 (with an R^2 of 0.94884). The PFO, while less accurate than the PSO model, still shows an accuracy of over 0.95 at pH_i 3, 7 and 1. Only at pH_i 5 and 9 the accuracy drops to below 0.95 at 0.89337 and 0.94656, respectively. The higher accuracy of the PSO model is again backed by the lower SSR values. Previous research on the use of adsorption models to describe desorption kinetics is few. Sarici-Ozdemir (2012) found that, for desorption of methylene blue from activated carbon, the PSO was most accurate with an R^2 of 0.934–0.978, depending on the methylene blue concentration. This was comparable to the PSO fitting of their adsorption, with an R^2 of 0.915–0.982. The PFO and Elovich models were significantly less accurate, with R^2 values of 0.756–0.929 and 0.800–0.875. Shirvani *et al.* (2007) tested five different models for the desorption of cadmium from the minerals palygorskite (consisting mostly of aluminium, magnesium and silicate) and sepiolite (consisting mostly of magnesium silicate). They found that, for fresh samples, the PSO model was most accurate (with an R^2 of 0.996 and 0.990, respectively) with the PFO (with an R^2 of 0.952 and 0.943) and Elovich (with an R^2 of 0.949 and 0.976) models slightly less accurate.

3.3. Influence of initial pH

Regeneration through NaCl, $FeCl_3$, $CaCl_2$ and $MgCl_2$ was tested at varying initial pH (pH_i). Figure 4 reveals the As(V) in the solution after desorption in the various media. Desorption in NaCl resulted in the highest As(V) concentration detected, at pH_i 12. The second highest desorption was found at pH_i 2, while desorption at pH_i 3 and 11 is among the lowest found among all metal chlorides. An even better indicator of regeneration efficiency is the adsorption following the regeneration, as also revealed in Figure 4. It is clear that the regeneration using NaCl solution, while displaying the highest As(V) concentration

Table 3 | Pseudo-first and pseudo-second order modeling of desorption kinetics at 20 g/L GAC loaded with 4.00 mg/g As(V), pH 3 ($FeCl_3$) or 11 ($CaCl_2$ and $MgCl_2$), 25 °C and a stirring speed of 150 rpm

		Pseudo-first order				Pseudo-second order			
	Ce_{exp}	Ce	k1	R^2	RSS	Ce	k2	R^2	RSS
In H_2O									
pH 3	20.25765	18.78496	0.03575	0.96227	14.33955	21.46236	0.002100	0.98577	5.40792
pH 5	9.44636	8.44051	0.06495	0.89337	8.28604	9.32527	0.009580	0.94884	3.97559
pH 7	9.16282	8.49137	0.09849	0.95556	3.57397	9.31541	0.013670	0.98961	0.83563
pH 9	10.82045	10.25372	0.04669	0.94656	5.97330	11.48449	0.005380	0.96786	3.59449
pH 11	35.41464	34.55503	0.01597	0.97186	30.28429	42.51215	0.000397	0.98528	15.84895
In $MeCl_x$									
Fe 3:1	36.86	36.15810	0.01853	0.88861	119.35876	38.85048	0.000716	0.92263	82.90305
Fe 6:1	30.34	22.45502	0.03470	0.71097	137.85227	25.20699	0.001790	0.83079	80.70293
Fe 12:1	15.96	14.77390	0.03379	0.93831	12.26398	16.11654	0.003060	0.98200	3.57933
Ca 3:1	31.42	33.88122	0.10394	0.97360	26.12921	35.81816	0.005080	0.97023	29.45931
Ca 6:1	36.92	38.71987	0.09757	0.98606	18.29914	41.16995	0.003950	0.98005	26.17656
Ca 12:1	41.62	41.96055	0.12305	0.97539	36.35085	44.48227	0.004630	0.99025	14.39595
Mg 3:1	39.44	38.11556	0.06835	0.97132	37.10551	41.10471	0.002600	0.99644	4.60167
Mg 6:1	45.54	41.47176	0.06648	0.95708	66.26658	45.07052	0.002200	0.99773	3.50322
Mg 12:1	55.34	47.38945	0.06749	0.91592	168.77535	51.75767	0.001900	0.98142	37.30122

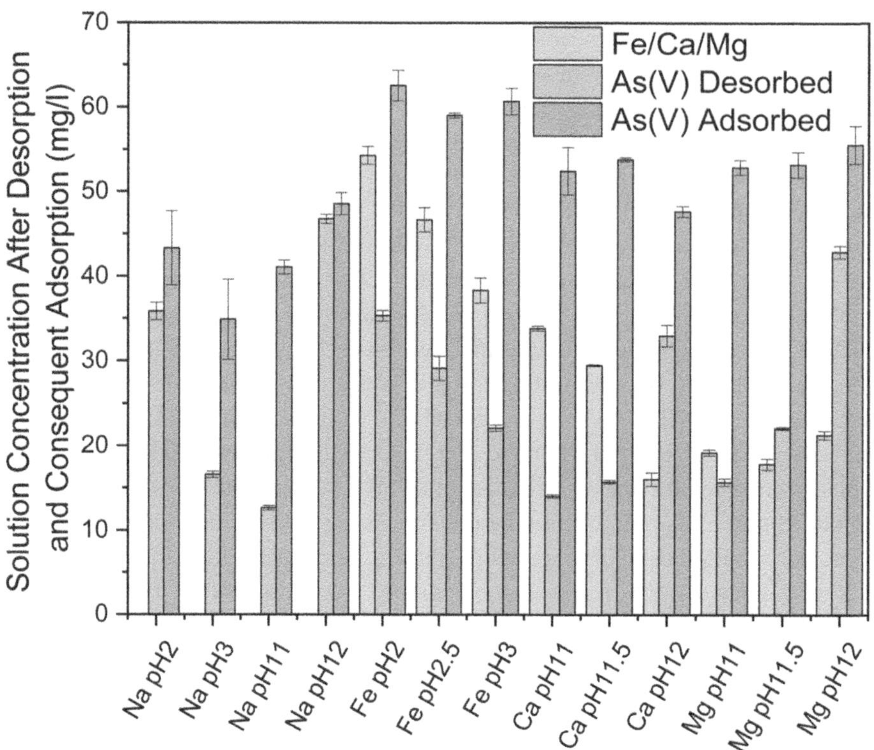

Figure 4 | The influence of metal chloride and pH on the concentration of As(V) and Fe/Ca/Mg post desorption and As(V) post adsorption at 20 g/L loaded GAC, 25 °C and a stirring speed of 150 rpm.

after desorption, showed the lowest subsequent adsorption efficiency out of all desorption tests. It is also revealed that, while As(V) desorption is higher at pH_i 3 than pH_i 11, subsequent adsorption is higher after regeneration at pH_i 11. Di Natale *et al.* (2013) found that desorption of arsenic from activated carbon was best in media of dissolved sodium salts (100% in NaOH, NaCl and $NaNO_3$ solution), but the pH_i was set at 8 for all experiments and subsequent adsorption was not tested. Therefore, while the results in this paper agree with the desorption data of Di Natale, the As(V) measured during desorption may not be the best indicator for regeneration efficiency. The pH_i of desorption affects the final pH of subsequent adsorption, as revealed in Figure 5. Whereas a desorption pH_i of 2 and 3 had little effect on the final pH of subsequent adsorption (with less than 0.1 difference), the difference between pH_i 3 and 11 was larger, with a final pH of 5.5 and 6.2, respectively. This change in final pH is explained by Noh & Schwarz (1990), who demonstrated that the surface acidity of activated carbon adjusts to that of the media it is submersed into, the degree of which depending on the difference in pH and the concentration of activated carbon. It is likely that, while As(V) desorption was higher at very high and low pH, the resulting pH is the cause for the diminished following adsorption.

When desorbing in $FeCl_3$ solution, a higher pH_i led to a lower concentration of As(V) desorbed. The concentration of Fe(III) in the solution is also revealed in Figure 4. The pH_i shows a clear effect on concentration, which is presumed to be through precipitation. The Fe(III) concentration decreases from 54.3 mg/L at pH_i 2 to 46.7 and 38.3 mg/L at pH_i 2.5 and 3, respectively. This corresponds to the decrease in As(V) detected at these pH_i, indicating that the lower As(V) concentration is due to precipitation. The pH after desorption, as listed in Figure 5, reveals that the pH goes up slightly, likely due the GAC surface acidity. For $FeCl_3$, an initial pH of 2.0, 2.5 and 3.0, led to a final pH of 2.1, 2.7 and 3.3, respectively. Robins (1987) examined and compared data of four studies to model the stability of scorodite, or ferric arsenate ($AsFeO_4$), by pH. Their model suggested that optimum removal took place at pH 4.8. There seems to be some disparity between optimum pH among papers, with the data of Nishimura & Tozawa (1978) showing this pH to be at 2. Ferric iron and arsenic exist in a variety of compounds such as angelellite ($Fe_4(AsO_4)_2(OH)_3$) and kamarizaite ($Fe_3(AsO_4)_2(OH)_3$). Majzlan *et al.* (2018) tested the precipitation of these compounds and scorodite and found that, while scorodite solubility was lowest in the range of pH $3.5 \sim 6.5$, for angellelite and kamarizaite the optimum was narrower, being close to 6.5. This seems to be

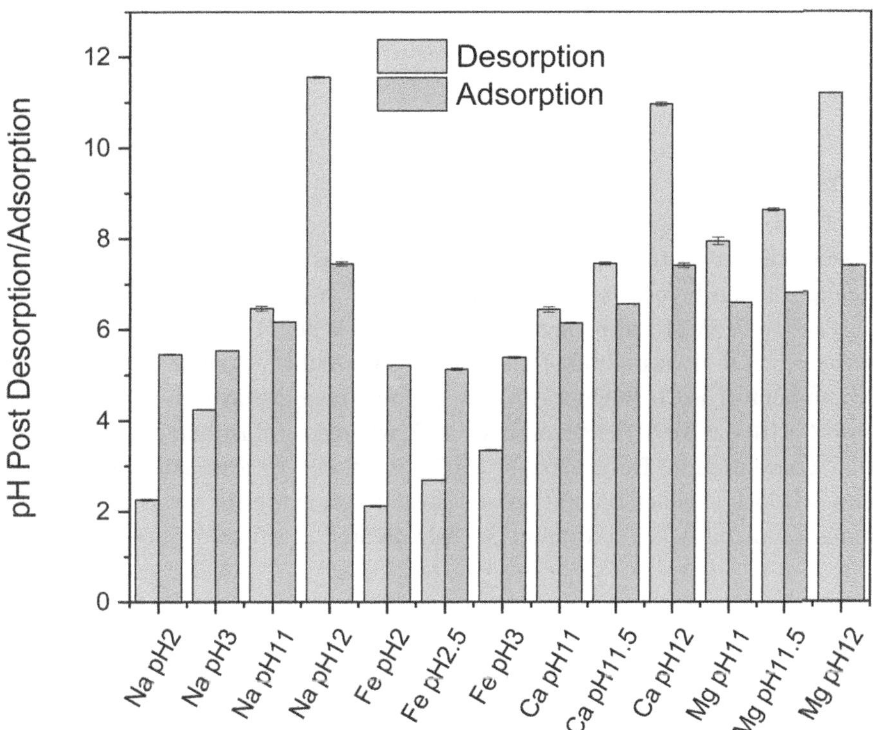

Figure 5 | The influence of metal chloride and pH on the final solution pH of desorption and adsorption at 20 g/L GAC loaded with 4.00 mg/g As(V), 25 °C and a stirring speed of 150 rpm.

in line with the data found in this paper, with the final pH of 3.3 leading to a lower detection of As(V) and Fe(III), even when dissolving in 10% HNO_3 solution. Subsequent adsorption of As(V) loaded GAC is highest following desorption in $FeCl_3$, according to Figure 4, and is highest at a pH_i of 2, followed by 3. The average adsorption is lowest at pH_i 2.5, which may be due to a limited overlap of desorption, which favors a pH removed from that of optimum adsorption pH, and precipitation, which was highest at pH_i 3. Considering the goal is regeneration of GAC while simultaneously removing As(V) from the solution this pH_i of 3 is considered optimal.

For desorption of As(V) in $CaCl_2$ aqueous solution Ca(II) concentration goes down with pH_i while As(V) concentration increases. This can be explained through the work of Bothe & Brown (1999b) who analyzed the phases of calcium arsenates at increasing pH. It was found that, as pH increases, the ratio of As(V) to Ca(II) in the predominant phase gradually decreases from 0.8 to 0.5. The relatively low ratio in this research this likely results in a lower fraction of bound As(V). Figure 5 lists the final pH after desorption in $CaCl_2$, where a pH_i of 11.0, 11.5 and 12.0 led to a final pH of 6.4, 7.5 and 11.0, respectively. When testing precipitation of various forms of calcium arsenates, Bothe & Brown (1999a) found that, at a pH of 11.2, the arsenic concentration had dropped to 3 mg/L and a pH of 12 further reduced this less than 0.5 mg/L. Zhang *et al.* (2015) similarly found that within their tested pH range (3–10) arsenic content in the precipitate increased with pH, where calcium transformed from gypsum ($CaSO_4$) to calcium arsenate ($Ca_3(AsO_4)_2$). This indicates that precipitation in this experiment likely takes place at the higher pH_i of 12, but is too low to be optimal. Subsequent adsorption seems to be optimal at pH_i 11.5, followed by pH_i 11, which lead to a final pH of 7.5 and 6.4, respectively. As(V) desorption, however, is slightly lower at pH_i 11. A pH_i of 12 leads to both the highest detected As(V) desorption and the lowest subsequent adsorption.

For desorption in Mg(II) solution, pH_i within this range does not have a major effect on Mg(II) concentration, while As(V) concentration increases with pH_i. A pH_i of 11.0, 11.5 and 12.0 led to a final pH of 7.9, 8.6 and 11.2, respectively. Park *et al.* (2010) found that for precipitation of As(V) with Mg, the pH for removal was optimal between 7.5 and 10.2, where arsenic existed in the form of $Mg_3(AsO_4)_2$. A higher pH will lead to precipitation of magnesium to $Mg(OH)_2$, which leads to a lower removal of arsenic. Li *et al.* (2019) tested precipitation of arsenic with magnesium in the presence of NH_4 and determined the optimum pH to be 9.5. The samples with pH_i 11.0 and 11.5 both fall within a favorable pH range, with 11.5 ending up closest to the optimum found by Li *et al.* Regeneration in $MgCl_2$ was slightly higher at pH_i 12, measuring by subsequent As(V) adsorption. The increase in adsorption is limited, however, compared to the As(V) desorbed during regeneration. Considering that a pH_i of 11 led to an only slightly lower regeneration efficiency, a better precipitation and requires less base addition it was deemed optimal.

3.4. Influence of molar ratio

The effect of molar ratio of iron, calcium and magnesium to arsenic was tested on the regeneration of As(V) loaded GAC. The concentration of As(V) when diluted in 10% HNO_3 solution and H_2O is revealed in Figure 6, in addition to adsorption of As(V) following regeneration. It can be observed that, when diluting in HNO_3, As(V) desorption increases with metal to arsenic ratio with the exception of Fe(III) at its highest molar ratio. When increasing the ratio from 6:1 to 12:1, a slight decrease in average As(V) desorption was found. This indicates that either maximum As(V) desorption is achieved at a ratio of 6:1, or the high amount of $FeCl_3$ might have been too much to dissolve in 10% HNO_3. Except for the ratio of 12:1, Fe(III) is superior to Ca(II) and Mg(II) in desorbing As(V), whereas at this ratio desorption in Fe(III) and Mg(II) solutions result in a similar concentration. Desorption in Ca(II) solution was least effective in the assisted desorption of As(V), except for the ratio of 0.75:1, at which desorption was close to that in Mg(II). It was found that there was a logarithmic relationship between molar ratio and As(V) desorption. A basic natural logarithmic model was fitted as follows:

$$Ce = a * ln(Ratio) + b \tag{7}$$

where Ce is the equilibrium concentration of As(V), the variable a is the multiplier of the ratio, Ratio is the molar ratio of Fe(III), Ca(II) and Mg(II) to As(V) and b is the concentration of As(V) at a ratio of 1. This model was fitted both for the complete set of ratios (labeled Log. All) and for the ratios from 0.75:1 to 6:1 (labeled Log. Excl.), due to discrepancy a a ratio of 12:1. It is observed in Figure 6 that, for desorption in Fe(III) aqueous solution, the fitting excluding the ratio of 12:1 showed a better fit. This is confirmed by the adjusted R^2 increasing from 0.90282 to 0.99992. For desorption in Ca(II) solution, a logarithmic fit is more accurate when using all ratios, with an R^2 of 0.96676 as opposed to 0.94387. For desorption in Mg(II) solution, the logarithmic fit was very accurate whether the 12:1 ratio was included or not, with an R^2 of 0.99820 and 0.99596, respectively.

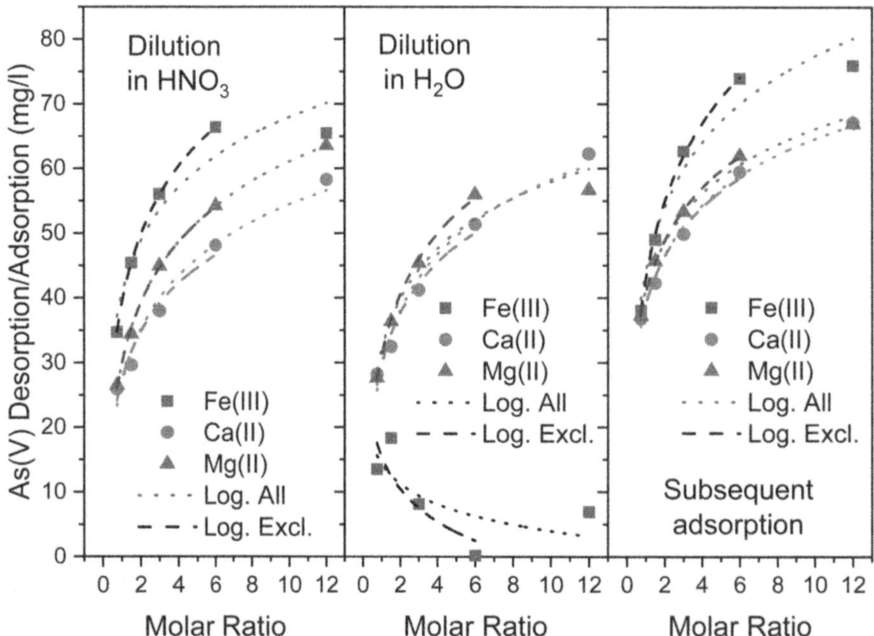

Figure 6 | The effect of molar ratio on the concentration of As(V) desorbed from GAC diluted in HNO$_3$ and H$_2$O and on subsequent adsorption of As(V) at 20 g/L GAC loaded with 4.00 mg/g As(V), pH 3 (FeCl$_3$) or 11 (CaCl$_2$ and MgCl$_2$), 25 °C and a stirring speed of 150 rpm.

The concentration of As(V) with samples diluted in H_2O is also revealed in Figure 6. The As(V) concentration when desorbed in $FeCl_3$ initially increases with molar ratio from 0.75:1 to 1.5:1, then decreases with molar ratio until 6:1. At this ratio an average concentration was found of 0.15 mg/L. When further increasing the ratio to 12:1, however, the As(V) concentration rises again. A higher molar ratio generally leads to a higher As(V) desorption, but also increases the amount of iron to coprecipitate As(V) with. This explains the initial rise in As(V) concentration, as the effect of increased As(V) desorption is stronger than that on the precipitation. At higher ratios the increase in precipitation is stronger than the increase in desorption. The mechanism behind the final rise in concentration at a ratio of 12:1 is unclear, however, as no increase was observed when diluted in HNO_3. The optimum molar ratio is 6:1 for both desorption of As(V) and its removal from the solution through precipitation, with near complete removal taking place (99.77% of the As(V) detected in HNO_3). The final pH for each molar ratio was measured between 2.6 and 3.2. No relation was found between this divergence in pH and the results. When desorbing in $MgCl_2$ the solubility of As(V) is unaffected by ratio in the range of 0.75:1 to 6:1, remaining nearly similar to the solubility in HNO_3. When further increasing the ratio from 6:1 to 12:1, however, solubility in water decreases, with the As(V) concentration increasing by only less than 1 mg/L. In comparison, the As(V) concentration increased by nearly 10 mg/L when diluting in HNO_3 solution when the molar ratio was increased from 6:1 to 12:1. The final pH went up greatly with molar ratio, being 6.9 at a ratio of 0.75:1 and 9.9 at 12:1. Li *et al.* (2019) showed that a pH of 9.5 was optimal, which is likely another cause for the increased precipitation at this highest ratio. When desorbing in $CaCl_2$ solution the molar ratio did not affect solubility of As(V). As(V) desorption in $CaCl_2$ solution behaved similarly whether dilution took place in HNO_3 solution or in H_2O. The final pH was relatively stable in a range of 6.8–7.1. A logarithmic fitting was tested again which was unsuitable for desorption in Fe(III) solution, reaching an adjusted R^2 of 0.35636 and 0.53932 for all ratios and excluding 12:1, respectively. This is expected, as the solubility in water was greatly diminished. For desorption in Ca(II) solution the logarithmic fitting showed good accuracy of 0.97097 and 0.95613, respectively, both of which are slightly higher as compared those found for the data in Figure 6. For desorption in Mg(II) solution the logarithmic model shows a good fitting in the range of ratio 0.75:1 to 6:1, at 0.99669, but is significantly worse when including the 12:1 ratio, at 0.94130. This is in contrast to the fitting for samples diluted in HNO_3, where an R^2 of over 0.99 was found. This fact reinforces the previous findings that, at this highest ratio and diluting in water, maximum solubility is reached.

The subsequent adsorption of As(V) by GAC regenerated in $FeCl_3$, $CaCl_2$ and $MgCl_2$ solutions is also displayed in Figure 6. While there is no significant difference between the metal chlorides at a molar ratio of 0.75:1, at all higher ratios, the regeneration in $FeCl_3$ aqueous solution is highest. The trend for As(V) desorbed in HNO_3 is reflected in subsequent As(V) adsorption, with the As(V) adsorbed 10% higher than the As(V) desorbed (with a maximum of 5.4% deviation). Similarly, a further increase in molar ratio from 6:1 to 12:1 has limited beneficial effect on the regeneration of As(V) loaded GAC, with the adsorption increasing from 73.9 to 75.9 mg/L (an increase of ~2.7%). This increase indicates that the decrease in As(V) concentration for desorption in $FeCl_3$ solution at a molar ratio of 12:1 is caused by precipitation even when diluting in 10% HNO_3 solution. Fresh GAC reached an adsorption of 80.2 mg/L, which means that regeneration in $FeCl_3$ aqueous solution led to a regeneration of 92.2% and 94.7% at molar ratios of 6:1 and 12:1, respectively, in comparison to adsorption of fresh GAC. From a ratio of 1.5:1 to 6:1 regeneration in $MgCl_2$ solution is superior to that in $CaCl_2$ but, at the ratios of 0.75:1 and 12:1, there is no significant difference. At Ca(II)/Mg(II):As(V) ratios of 12:1, regeneration in $CaCl_2$ and $MgCl_2$ led to a regeneration efficiency of 83.6% and 83.4%, respectively. Comparing the three metal chlorides, $FeCl_3$ is superior to $CaCl_2$ and $MgCl_2$ in As(V) desorption, As(V) precipitation and As(V) adsorption following regeneration. A logarithmic model was fitted to the adsorption data in Figure 6, which was comparable to the fittings for desorption of As(V) when diluted in HNO_3 solution. Regeneration in $FeCl_3$ solution led to an adjusted R^2 0.94458 when fitting all ratios and 0.99739 with the exclusion of 12:1, both of which were higher than the R^2 for desorption. Adsorption following regeneration in $CaCl_2$ was very accurate at 0.99034 when fitting all ratios and dropped slightly to 0.97955 with the exclusion of 12:1, an increase in accuracy over the data for desorption. Adsorption following regeneration in $MgCl_2$ was most accurate regardless of whether or not the ratio of 12:1 was included, with an R^2 of 0.99090 and 0.99917, respectively.

While other literature investigating regeneration of activated carbon loaded with heavy metals through coagulation was not found, other methods have been tested. Hamdaoui *et al.* (2005) employed ultrasound to regenerate granular activated carbon loaded with Hg(II), Cr(VI), Cu(II) and Mn(II), reaching regeneration efficiencies of 24%, 35%, 37% and 43%, respectively. The same technique was used by Jing *et al.* (2011) for regeneration of Cr(VI) from powdered activated carbon which reached an efficiency of 29%. When eluting Pt and Pd from activated carbon using NaCl solution, Snyders *et al.* (2015) found a high desorption of 99%, but this required a temperature of 95 °C. At 60 °C, efficiencies were markedly lower at 64% and 68%, respectively. Removal of Au was significantly lower, reaching ~32% at 95 °C and being nonexistent at 60 °C. When adsorption is exothermic, higher temperature will aid in desorption (Van Deventer & Van der Merwe 1994). This higher temperature, while increasing regeneration efficiency, leads to a significant increase in operating costs. Whereas the previously mentioned research of Di Natale *et al.* (2013) reached complete desorption of As and a regenerated adsorption capacity of 93%, this was at a relatively low loading of 0.1 mg/g. In comparison, the arsenic removal by electrocoagulation in this study displays high removal while taking place at room temperature and high loading. For regeneration of cadmium loaded activated carbon, made from highly ligno-cellulosic jute stick, using 0.1 mol/L HNO_3, Ghosh *et al.* (2021) found that 95.4% desorption of cadmium took place. This led to a subsequent adsorption of 96.3% of the original 73,53 mg/g. The highest regeneration of 94.4% (measured as subsequent adsorption) found in this research is very comparable, proving the feasibility of coagulation in the regeneration of activated carbon.

Elemental analysis was performed by XRF, as listed in Table 4. Surface arsenic is found to increase to 6.5% after loading and decrease to 1.4–1.6% after regeneration with the exception of Fe 6:1, which is slightly higher at 2.5%. This could be due to the relatively high iron content, which is commonly doped onto AC to improve arsenic adsorption (Lorenzen *et al.* 1995). Regeneration with $FeCl_3$ leads to higher concentrations than regeneration with calcium and magnesium chloride. The same is true for calcium after regeneration with $CaCl_2$, with the observed 2.1% the highest among samples. This is not observed for magnesium after regeneration with $MgCl_2$, however, with the detected 22.0% being among the lowest found between samples. This is likely due to the magnesium content on the GAC already being among the highest of elements measured, making it more difficult to additional magnesium to bind to the surface. Chlorine concentration decreased after loading with As(V) and was found to be highest after regeneration with $FeCl_3$ at a molar ratio of 3:1 and 12:1. After regeneration it was found to be the lowest after regeneration with $MgCl_2$, which corresponds to the low adsorption found for magnesium. For future experiments, the use of EDX, CHNS or XPS may provide additional insights into deposition on the GAC.

Surface area and pore diameter were tested by BET method, also listed in Table 4. It was found that surface area increases slightly through regeneration with $FeCl_3$ and $CaCl_2$ and more effectively with $MgCl_2$. In contrast, the pore diameter decreases slightly during the regeneration, most visibly for $CaCl_2$. The effect of this seems low, however, with the regeneration of GAC as displayed in Figure 7 equally effective when using $CaCl_2$ and $MgCl_2$.

SEM analysis was performed but the regeneration solution showed no influence on the observed GAC surface.

3.5. Regeneration kinetics

The desorption was modeled in $FeCl_3$, $CaCl_2$ and $MgCl_2$ aqueous solution through PFO and PSO fittings, as revealed in Figure 7. Unlike the data in Figure 6, for desorption in $FeCl_3$ solution a higher molar ratio leads to a lower As(V) detection, despite the samples being similarly diluted in 10% HNO_3. This is likely caused by the higher reaction volume leading to insufficient mobilization of precipitate. Table 3 reveals the fitting data for the PFO and PSO models. In all cases, the PSO was more accurate in describing the desorption of As(V) from GAC in $FeCl_3$ solution. Due to precipitation taking place the accuracy is low, however, with only desorption at a molar ratio of 12:1 leading to an R^2 over 0.95. This is confirmed by the SSR data, which are larger for the PFO model. For desorption in $CaCl_2$ solution, as in Figure 6, a molar ratio leads to an increased

Table 4 | Surface analysis of activated carbon through XRF and BET

XRF	Mg %	Al %	Si %	S %	Cl %	Ca %	Ti %	Fe %	As %	Surface A. m²/g	Pore diam. nm
Fresh	20.0%	10.4%	34.7%	8.1%	17.9%	1.3%	1.5%	6.1%	99.8[a]	946.56	2.2222
Loaded	22.6%	11.2%	33.3%	8.8%	11.2%	1.0%	1.4%	4.0%	6.5%	–	–
Fe 3:1	21.8%	9.8%	27.8%	8.2%	23.7%	0.9%	1.1%	5.1%	1.4%	–	–
Fe 6:1	26.9%	7.5%	22.5%	4.4%	16.0%	0.7%	1.0%	18.6%	2.5%	–	–
Fe 12:1	25.8%	8.8%	23.5%	6.8%	22.7%	0.8%	1.2%	8.9%	1.4%	956.09	2.1748
Ca 12:1	22.8%	10.4%	30.1%	8.5%	19.9%	2.1%	1.5%	3.1%	1.6%	978.98	2.0943
Mg 12:1	22.0%	10.8%	33.9%	8.3%	15.9%	1.1%	1.4%	5.1%	1.4%	1,080.00	2.1739

[a]low detection, measured in ppm.

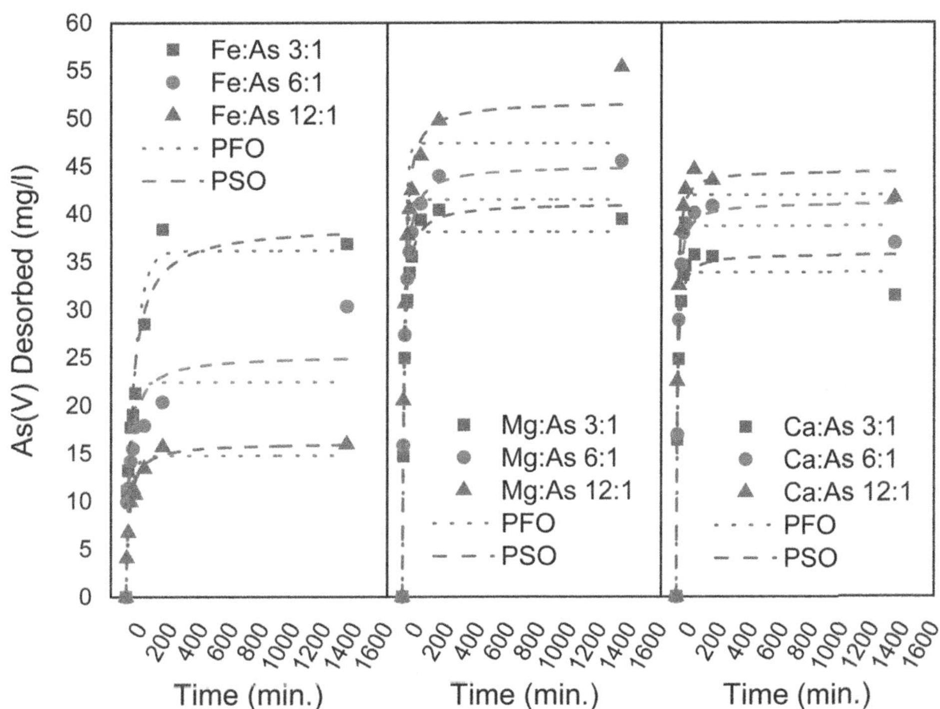

Figure 7 | Pseudo-first and pseudo-second order model comparison for As(V) desorption from GAC in FeCl$_3$, CaCl$_2$ and MgCl$_2$ aqueous solution at molar ratios of 3:1, 6:1 and 12:1 at 20 g/l GAC loaded with 4.00 mg/g As(V), pH 3 (FeCl$_3$) or 11 (CaCl$_2$ and MgCl$_2$), 25 °C and a stirring speed of 150 rpm.

As(V) desorption but the As(V) concentrations detected were lower in comparison. This is likely also caused by the use of 250 mL flasks. Both the PFO and PSO fit the data with good accuracy ($R^2 > 0.95$) with the PFO being slightly superior at 3:1 and 6:1 ratio reaching an R^2 of 0.97360 and 0.98606, respectively, and the PSO being superior at a 12:1 ratio, with an R^2 of 0.99025. The SSR values reveal that the PFO model was a slightly better fit at ratios of 3:1 and 6:1, whereas the PSO was a better fit at a ratio of 12:1. For desorption in MgCl$_2$ solution, As(V) desorption increases with molar ratio but is lower than previously found, similar to desorption in CaCl$_2$ solution. While the PFO fitting is lower than that found for CaCl$_2$, the PSO fitting is the highest found, exceeding 0.98 at a ratio of 12:1 and 0.99 at 3:1 and 6:1. This is confirmed by the lower values SSR values. These results match those in Section 3.2, with the PSO generally outperforming that of the PFO. This also matches the literature discussed in that section, with the PSO fitting for desorption in dH$_2$O, CaCl$_2$ and MgCl$_2$ in this paper being more accurate than those by Sarici-Ozdemir (2012) but lower than those by Shirvani *et al.* (2007). The fitting for desorption in FeCl$_3$ solution is significantly lower, due to precipitation taking place.

4. CONCLUSIONS

The increase in adsorption capacity of As(V) on granular activated carbon (GAC) went down when increasing the initial As(V) concentration from 150 to 200 m/L, indicating that the GAC was near completely satisfied. The pseudo-second order (PSO) was more accurate than the pseudo-first order (PFO). The Freundlich model was most accurate followed by the Temkin model and finally the Langmuir model. Desorption of As(V) loaded GAC in dH$_2$O was highest at pH 11 and followed by pH 3, removed from the previously found optimum pH for adsorption of 6. PSO modeling was most accurate for desorption. Desorption of As(V) from GAC was tested. Despite NaCl leading to the highest desorbed As(V) detected it displayed the lowest As(V) adsorption following regeneration. Regeneration in FeCl$_3$ showed the highest subsequent As(V) adsorption, with the largest precipitation taking place at pH 3.0. Regeneration in CaCl$_2$ and MgCl$_2$ was both deemed optimal at pH 11. Regeneration of GAC and precipitation of As(V) generally increased with the molar ratio of Fe(III)/Ca(II)/Mg(II) to As(V). A logarithmic relationship between molar ratio and As(V) desorption and subsequent adsorption was found. Subsequent As(V) adsorption was highest in FeCl$_3$ and lowest in CaCl$_2$. Kinetic modeling was inaccurate for desorption in

FeCl$_3$. For desorption in CaCl$_2$ the PFO was a marginally better fit at the molar ratios of 3:1 and 6:1, but the PSO was more accurate at 12:1. Desorption in MgCl$_2$ was best described with the PSO at every ratio. In conclusion, regeneration of As(V) loaded GAC was best achieved in FeCl$_3$ aqueous solution at a pH of 3 and a molar ratio of 12:1 when subsequent adsorption is the only goal. However, a molar ratio of 6:1 was nearly as effective while requiring half the amount of FeCl$_3$ and removing As(V) more efficiently through precipitation. This precipitation, however, results in kinetic modeling becoming more difficult to achieve in comparison to that in CaCl$_2$ and MgCl$_2$.

AUTHOR CONTRIBUTIONS

N. M. Moed: conceptualization; data curation; formal analysis; investigation; methodology; software; validation; visualization; writing – original draft & editing; Y. Ku: funding acquisition; project administration; resources; supervision

DATA AVAILABILITY STATEMENT

All relevant data are included in the paper or its Supplementary Information.

CONFLICT OF INTEREST

The authors declare there is no conflict.

REFERENCES

Al-Ghouti, M. A. & Da'ana, D. A. 2020 Guidelines for the use and interpretation of adsorption isotherm models: a review. *Journal of Hazardous Materials* **393**, 122383. https://doi.org/10.1016/j.jhazmat.2020.122383.

Anderson, M. A., Ferguson, J. F. & Gavis, J. 1976 Arsenate adsorption on amorphous aluminum hydroxide. *Journal of Colloid And Interface Science* **54** (3), 391–399.

Ania, C. O., Parra, J. B., Menéndez, J. A. & Pis, J. J. 2005 Effect of microwave and conventional regeneration on the microporous and mesoporous network and on the adsorptive capacity of activated carbons. *Microporous and Mesoporous Materials* **85** (1–2), 7–15.

Armstrong, C. W., Stroube, R. B., Miller, G. B., Rubio, T. & Siudyla, E. A. 1984 Outbreak of fatal arsenic poisoning caused by contaminated drinking water. *Archives of Environmental Health* **39** (4), 276–279.

Ashrafi, M., Bagherian, G. & Goudarzi, N. 2018 Removal of brilliant green and crystal violet from mono-and Bi-component aqueous solutions using NaOH-modified walnut shell. *Iranian Chemical Society Analytical and Bioanalytical Chemistry Research* **5** (1), 95–114.

Banerjee, M., Bhattacharjee, P. & Giri, A. K. 2011 Arsenic-induced cancers: a review with special reference to gene, environment and their interaction. *Genes and Environment* **33** (4), 128–140.

Batool, F., Akbar, J., Iqbal, S., Noreen, S. & Bukhari, S. N. A. 2018 Study of isothermal, kinetic, and thermodynamic parameters for adsorption of cadmium: an overview of linear and nonlinear approach and error analysis. *Bioinorganic Chemistry and Applications*. https://doi.org/10.1155/2018/3463724.

Bothe, J. V. & Brown, P. W. 1999a Arsenic immobilization by calcium arsenate formation. *Environmental Science and Technology* **33** (21), 3806–3811.

Bothe, J. V. & Brown, P. W. 1999b The stabilities of calcium arsenates at 23 $\pm$ 1 °C. *Journal of Hazardous Materials* **69** (2), 197–207.

Davoodi, S., Dahrazma, B., Goudarzi, N. & Gorji, H. G. 2019 Adsorptive removal of azithromycin from aqueous solutions using raw and saponin-modified nano diatomite. *Water Science and Technology* **80** (5), 939–949. https://doi.org/10.2166/wst.2019.337.

Di Natale, F., Erto, A. & Lancia, A. 2013 Desorption of arsenic from exhaust activated carbons used for water purification. *Journal of Hazardous Materials* **260**, 451–458. http://dx.doi.org/10.1016/j.jhazmat.2013.05.055.

Fouda-Mbanga, B. G., Prabakaran, E. & Pillay, K. 2021 Carbohydrate biopolymers, lignin based adsorbents for removal of heavy metals (Cd^{2+}, Pb^{2+}, Zn^{2+}) from wastewater, regeneration and reuse for spent adsorbents including latent fingerprint detection: a review. *Biotechnology Reports* **30**, 1–16. https://doi.org/10.1016/j.btre.2021.e00609

Ge, H. & Fan, X. 2011 Adsorption of Pb^{2+} and Cd^{2+} onto a novel activated carbon-chitosan complex. *Chemical Engineering and Technology* **34** (10), 1745–1752. https://doi.org/10.1002/ceat.201000182.

Ghosh, R. K., Ray, D. P., Chakraborty, S., Saha, B., Manna, K., Tewari, A. & Sarkar, S. 2021 Cadmium removal from aqueous medium by jute stick activated carbon using response surface methodology: factor optimisation, equilibrium, and regeneration. *International Journal of Environmental Analytical Chemistry* **101** (14), 2171–2188. https://doi.org/10.1080/03067319.2019.1700964.

Gong, X., Li, W., Zhang, D., Fan, W. & Zhang, X. 2015 Adsorption of arsenic from micro-polluted water by an innovative coal-based mesoporous activated carbon in the presence of co-existing ions. *International Biodeterioration & Biodegradation* 1–9. http://dx.doi.org/10.1016/j.ibiod.2015.01.007.

Gupta, S. K. & Chen, K. Y. 1978 Arsenic removal by adsorption. *Water Pollution Control Federation* **50** (3), 493–506.

Halder, D., Lin, J., Essilfie-Dughan, J., Das, S., Robertson, J. & Hendry, M. J. 2018 Implications of the iron(II/III)-arsenic ratio on the precipitation of iron-arsenic minerals from pH 2.5 to 10.5. *Applied Geochemistry* **98**, 367–376.

Hamdaoui, O., Djeribi, R. & Naffrechoux, E. 2005 Desorption of metal ions from activated carbon in the presence of ultrasound. *Industrial and Engineering Chemistry Research* **44** (13), 4737–4744. https://doi.org/10.1021/ie048851t.

Hassan, A. F., Abdel-mohsen, A. M. & Elhadidy, H. 2014 Adsorption of arsenic by activated carbon, calcium alginate and their composite beads. *International Journal of Biological Macromolecules* **68**, 125–130. http://dx.doi.org/10.1016/j.ijbiomac.2014.04.006.

Herbert, A. & Kumar, U. 2021 Removal of acid blue 93 dye from aqueous media using an agro-industrial waste as bioadsorbent: an OPAT and DoE approach. *International Journal of Environmental Analytical Chemistry*. https://doi.org/10.1080/03067319.2021.1940987

Ho, Y. 2006 Review of second-order models for adsorption systems. *Journal of Hazardous Materials* **136**, 681–689.

Ho, Y. S. & Mckay, G. 1999 Pseudo-second order model for sorption processes. *Process Biochemistry* **34**, 451–465.

Hong, H.-Y., Moed, N. M., Ku, Y. & Lee, H.-Y. 2020 Ultrasonic regeneration studies on activated carbon loaded with isopropyl alcohol. *Applied Sciences (Switzerland)* **10** (21), 1–13.

Jing, G., Zhou, Z., Song, L. & Dong, M. 2011 Ultrasound enhanced adsorption and desorption of chromium (VI) on activated carbon and polymeric resin. *Desalination* **279**, 423–427. https://doi.org/10.1016/j.desal.2011.06.001.

Johnson, R. D. & Arnold, F. H. 1995 The temkin isotherm describes heterogeneous protein adsorption. *Biochimica et Biophysica Acta (BBA)/Protein Structure and Molecular* **1247** (2), 293–297.

Johnston, R. B. & Singer, P. C. 2007 Solubility of symplesite (Ferrous arsenate): implications for reduced groundwaters and other geochemical environments. *Soil Science Society of America Journal* **71** (1), 101–107.

Koohzad, E., Jafari, D. & Esmaeili, H. 2019 Adsorption of lead and arsenic ions from aqueous solution by activated carbon prepared from tamarix leaves. *ChemistrySelect* **4**, 12356–12367.

Lagergren, S. Y. 1898 Zur theorie der sogenannten adsorption gelöster Stoffe, Kungliga Svenska Vetenskapsakad. *Handlingar* **24**, 1–39.

Langmuir, I. 1916 The constitution and fundamental properties of solids and liquids. Part II.–Liquids. *Journal of the American Chemical Society* **38** (11), 2221–2295.

Larasati, A., Fowler, G. D. & Graham, N. J. D. 2021 Insights into chemical regeneration of activated carbon for water treatment. *In Journal of Environmental Chemical Engineering* **9** (4), 1–11. https://doi.org/10.1016/j.jece.2021.105555.

Li, T., Zhang, Y., Zhang, B., Chang, K., Jiao, F. & Qin, W. 2019 Arsenic(V) removal from enargite leach solutions by precipitation of magnesium ammonium arsenate. *Separation Science and Technology (Philadelphia)* **54** (11), 1862–1870. https://doi.org/10.1080/01496395.2018.1538245.

Lorenzen, L., van Deventer, J. S. J. & Landi, W. M. 1995 Factors affecting the mechanism of the adsorption of arsenic species on activated carbon. *Minerals Engineering* **8** (4–5), 557–569.

Majzlan, J., Nielsen, U. G., Dachs, E., Benisek, A., Drahota, P., Kolitsch, U., Herrmann, J., Bolanz, R. & Ševko, M. 2018 Thermodynamic properties of mansfieldite (AlAsO$_4$·2H$_2$O), angelellite (Fe4(AsO$_4$)2O$_3$) and kamarizaite (Fe$_3$(AsO$_4$)$_2$(OH)$_3$ ·3H$_2$O). *Mineralogical Magazine* **82** (6), 1333–1354.

Mandal, B. K. & Suzuki, K. T. 2002 Arsenic round the world: a review. *Talanta* **58**, 201–235.

Martin, R. J. & Ng, W. J. 1985 Chemical regeneration of exhausted activated carbon–II. *Water Research* **19** (12), 1527–1535.

McLaughlin, H. S. 1995 Regenerate activated carbon using organic solvents. *Chemical Engineering Progress* **91** (7), 45–53.

Milton, A. H., Hasan, Z., Rashman, A. & Rahman, M. 2001 Arsenic poisoning and respiratory effects in Bangladesh. *Journal of Occupational Health* **43**, 136–140.

Moed, N. M. & Ku, Y. 2022 The effect of Fe(II), Fe(III), Al(III), Ca(II) and Mg(II) on electrocoagulation of As(V). *Water (Switzerland)* **14** (2), 1–16. https://doi.org/10.3390/w14020215.

Nidheesh, P. V. & Anantha Singh, T. S. 2017 Arsenic removal by electrocoagulation process : recent trends and removal mechanism. *Chemosphere* **181**, 418–432. http://dx.doi.org/10.1016/j.chemosphere.2017.04.082.

Nishimura, T. & Tozawa, K. 1978 On the solubility products of ferric, calcium and magnesium arsenates. *Bulletin of the Research Institute of Mineral Dressing and Metallurgy (Tohoku University) (in Japanese)* **34**, 20–26.

Noh, J. S. & Schwarz, J. A. 1990 Effect of HNO3 treatment on the surface acidity of activated carbons. *Carbon* **28** (5), 675–682.

Palfy, P., Vircikova, E. & Molnar, L. 1999 Processing of arsenic waste by precipitation and solidification. *Waste Management* **19** (1), 55–59.

Pantuzzo, F. L., Santos, L. R. G. & Ciminelli, V. S. T. 2014 Solubility-product constant of an amorphous aluminum-arsenate phase (AlAsO4·3.5H2O) at 25 C. *Hydrometallurgy* **144–145**, 63–68. http://dx.doi.org/10.1016/j.hydromet.2014.01.001.

Park, Y. Y., Tran, T., Lee, Y. H., Nam, Y. I., Senanayake, G. & Kim, M. J. 2010 Selective removal of arsenic(V) from a molybdate plant liquor by precipitation of magnesium arsenate. *Hydrometallurgy* **104** (2), 290–297. http://dx.doi.org/10.1016/j.hydromet.2010.07.002

Pierce, M. L. & Moore, C. B. 1982 Adsorption of arsenite and arsenate on amorphous iron hydroxide. *Water Research* **16**, 1247–1253.

Robins, G. R. 1987 Solubility and stability of scorodite, FeAsO4.2H2O: discussion. *American Mineralogist* **72** (7–8), 849–851.

Salvador, F. & Jiménez, C. S. 1996 A New method for regenerating activated carbon by thermal desorption with liquid water under subcritical conditions. *Carbon* **34** (4), 511–516.

San Miguel, G., Lambert, S. D. & Graham, N. J. D. 2001 The regeneration of field-spent granular-activated carbons. *Water Research* **35** (11), 2740–2748.

Sarici-Ozdemir, C. 2012 Adsorption and desorption kinetics behaviour of methylene blue onto activated carbon. *Physicochemical Problems of Mineral Processing* **48** (2), 441–454.

Shirvani, M., Shariatmadari, H. & Kalbasi, M. 2007 Kinetics of cadmium desorption from fibrous silicate clay minerals: influence of organic ligands and aging. *Applied Clay Science* **37**, 175–184.

Snyders, C. A., Bradshaw, S. M., Akdogan, G. & Eksteen, J. J. 2015 Factors affecting the elution of Pt, Pd and Au cyanide from activated carbon. *Minerals Engineering* **80**, 14–24. https://doi.org/10.1016/j.mineng.2015.06.013.

Sun, Z., Yu, Y., Pang, S. & Du, D. 2013 Manganese-modified activated carbon fiber (Mn-ACF): novel efficient adsorbent for arsenic. *Applied Surface Science* **284**, 100–106. http://dx.doi.org/10.1016/j.apsusc.2013.07.031.

Van Deventer, J. S. J. & Van Der Merwe, P. F. 1994 The mechanism of elution of gold cyanide from activated carbon. *Metallurgical and Materials Transactions B* **25** (6), 829–838. https://doi.org/10.1007/BF02662765.

Waypa, J. J., Elimelech, M. & Hering, J. G. 1997 Arsenic removal by RO and NF membranes. *Journal/American Water Works Association* **89** (10), 102–114.

Yao, S., Liu, Z. & Shi, Z. 2014 Arsenic removal from aqueous solutions by adsorption onto iron oxide/Activated carbon magnetic composite. *Journal of Environmental Health Science and Engineering* **12** (58), 6–13.

Zhang, D., Yuan, Z., Wang, S., Jia, Y. & Demopoulos, G. P. 2015 Incorporation of arsenic into gypsum: relevant to arsenic removal and immobilization process in hydrometallurgical industry. *Journal of Hazardous Materials* **300**, 272–280. http://dx.doi.org/10.1016/j.jhazmat.2015.07.015.

First received 1 October 2021; accepted in revised form 5 August 2022. Available online 11 August 2022

doi: 10.2166/wst.2022.041

An improved primary wastewater treatment system for a slaughterhouse industry: a full-scale experience

Akshay D. Shende[a,b], Swati Dhenkula[a], Neti Nageswara Rao[a] and Girish R. Pophali[a,*]

[a] Wastewater Technology Division, CSIR-National Environmental Engineering Research Institute, Nagpur 440020, India
[b] Environmental Technology Division, CSIR-National Institute for Interdisciplinary Science and Technology, Thiruvananthapuram 695019, India
*Corresponding author. E-mail: gr_pophali@neeri.res.in, girishrpophali@gmail.com

ABSTRACT

The effluent streams from individual slaughtering operations were segregated based on the degree of similarity and were treated separately. The wastewater from lairage and paunch sections was dominant in suspended solids (SS: 6,000–25,000 mg/L) and was separated using a hydrasieve (500 µm) and externally fed rotary drum filter (EFRDF, 200 µm), respectively. The SS removal efficiency of the hydrasieve and EFRDF was 75% and 55%, respectively, and remaining solids were removed through a primary clarifier. The fats, oils and grease (FOG: 12,000–35,000 mg/L) containing streams from the hide fleshing, rendering, intestine, and tripe washing were routed through a skimming tank. The SS and FOG removal efficiencies through the skimming tank were 75% and 90%, respectively. Any FOG remaining after the skimming tank was removed using dissolved air flotation which achieved 95% FOG removal. In addition, the efficiency of chemical oxygen demand removal through the primary treatment system was more than 80%. The effluent obtained after primary treatment was SS and FOG ≤ 200 and 100 mg/L. The segregation of streams and their separate treatment offered benefits such as resource recovery, reduced waste load on downstream secondary treatment and overall ease in slaughterhouse wastewater treatment.

Key words: primary treatment, resource recovery, segregation of streams, slaughterhouse industry, wastewater

HIGHLIGHTS

- Segregation of streams and the removal of dung solids (TSS) and fats, oils, and grease (FOG) through multi-intervention approach is highly efficient.
- Hydrasieve-drum filter-primary clarifier ensures TSS removal to less than 200 mg/L.
- Skimming tank–DAF ensures FOG removal between 80 and 100 mg/L.
- Segregation of streams based on the nature of similarity and separate treatment is key to managing the slaughterhouse wastewater.

INTRODUCTION

The efficient primary wastewater treatment is the most crucial step in determining the overall success of any industrial effluent treatment plant (ETP). The primary wastewater treatment has two basic objectives: (1) separation of suspended solids (SS) and reduction of biochemical oxygen demand (BOD); and (2) utilization of separated materials for making commercial products (USEPA 2002). One of the necessary steps while dealing with slaughterhouse wastewater is its effective primary wastewater treatment. SS and fats, oils, and grease (FOG) contribute a major burden in slaughterhouse wastewater. Several researchers have reported the problems encountered due to the ineffective primary treatment of slaughterhouse wastewater. Anaerobic digestion is a versatile process for treating wastewater with a high organic concentration. Still, the accumulation of SS and FOG may lead to loss of sludge (biomass washout), limit the chances of operating at high organic loading, and deteriorate the specific methanogenic activity in upflow anaerobic sludge blanket (UASB) and anaerobic suspended biomass reactors. Chocking and clogging due to SS hinder the performance of anaerobic fixed film reactors treating slaughterhouse wastewater (Sunder & Satyanarayan 2013). Miranda *et al.* (2005) stated that an oil and grease/chemical oxygen demand (COD) ratio above 20% in the influent causes low-performance efficiency, biomass washout and a failure of the UASB system. Thus, the performance of these anaerobic technologies relies on the effective removal of SS and FOG. High FOG content in wastewater has an adverse effect because of its insoluble nature, slowing down the degradation rate. FOG may also prove problematic because of its capacity to form scum and coat surfaces. Moreover, SS and FOG are the prime reasons for the fouling of membranes used in the filtration process (Ghaffour 2004). Effective primary treatment of slaughterhouse wastewater is therefore essential to ensure performance of downstream secondary biological treatment systems.

The target pollutants in the primary treatment of slaughterhouse wastewaters are SS and FOG. Table 1 presents the summary of the various solid–liquid separation equipment used to remove SS from the effluent. Fleming & Macalpine (2003) evaluated the screw press for swine manure treatment using a screen with an opening of 0.25 mm with axial wire. It was found that because of the low solid content in the influent, it was difficult to develop a good liquid seal in the solid discharge; as a result, the dry matter (DM) percentage (%) in the effluent was more (0.92%) than the influent (0.84%). A vibratory screen with a 0.212-mm screen opening was used for the influent from the dairy farm, which had a DM content of less than 1.85%. The DM removal efficiency of nearly 11.89% was achieved. However, flow rates had to be reduced over time as the solid level rose while operating the vibratory screen (Fleming & Macalpine 2003). Fernandes *et al.* (1988) developed continuous belt microscreening to manage swine wastewater using a belt made of a filter medium with a 100-µm opening and involved continuous cleaning by airflow. It was operated at a hydraulic loading rate (HLR) of 0.145 m^3/m^2·min. The efficiency of the continuous belt microscreening unit increased with an increase in initial SS concentration in the influent slurry. The DM removal efficiency of 53% for the initial DM of 4.7% was increased to 60% for the initial DM of 8% in the influent slurry.

Gooch *et al.* (2005) assessed the performance of the screw press for three different dairy farms and found that its performance increased as the input solid concentration in the influent stream increased. DM percentage removal efficiency was 50.5% when the initial DM% in the influent was 9.96%, and it reduced to 39% when the DM% in the influent stream was nearly 8.32%. Although the screw press performed satisfactorily with a higher DM% in the influent stream, it is essential to have a screen opening of appropriate size. Gooch *et al.* (2005) found only 0.05% DM removal efficiency with the screen opening of 2.25 mm while operating a screw press for one of the dairy farms with an average flow of 12.75 m^3/h. Giorgia (2013) assessed the dewatering of anaerobically digested livestock slurry using a decanting centrifuge. The initial solid concentration in the influent slurry was 5%, and that of the outlet was between 2.3% and 2.4%. A solid removal efficiency of 52–54% was gained for a feed flow rate of 4–6 m^3/h. The solid concentration in separated solids was 21.7–29%. The electricity consumption of the decanting centrifuge was 1 kWh/m^3, considerably more than other solid–liquid separation technologies. The author also demonstrated that the decanting centrifuge's performance could be further enhanced by adding polymers in the influent slurry. Gilkinson & Frost (2007) also carried out a similar study to check the vibratory screen performance and decanting centrifuge to handle cattle slurry. The vibratory screen achieved solid removal efficiency of 45% compared to the centrifuge which achieved 70% solids removal efficiency. However, the power requirement for the vibratory screen was only 0.075 kWh/m^3 and that for centrifuge was nearly 1.5 kWh/m^3.

It was observed that regardless of the different solid–liquid separation technologies used to manage livestock slurry, DM removal efficiency was around 50%. At the same time, it is equally important to consider the power required to run these separation systems. Power requirement for the decanter centrifuge was higher (1.5 kWh/m^3) (Gilkinson & Frost 2007), and for the screw press, it was 0.239–0.625 kWh/m^3 (Fleming & Macalpine 2003). One of the crucial parameters that decide the success of the solid–liquid separation systems is the pore size of the screen. As shown by Fernandes *et al.* (1988) using a continuous belt microfiltration unit with a filter fabric with an opening size of 100 µm, DM removal efficiency of 53–60% was achieved. In a particle-size distribution study of cattle slurry carried out by Salehion *et al.* (2013), it was found that a 0.7-mm sieve retained 87.6% DM. Wright (2018) found that 78% of the total mass was larger than 0.63 mm, and 51% of the total mass was larger than 2.5 mm; as a result, they used a simple inclined screen with an opening of 2.5 mm and achieved SS removal efficiency of 51–58%. It is imperative to mention here that the SS removal efficiency in the case of Wright (2018) can be further improved by using screens with smaller openings. In the market, inclined and rotary drum screens are available with opening sizes starting from 100 to 1,000 µm.

Moreover, the size and composition of particles from livestock/dairy farms/lairage and paunch sections vary with animal species, diet and period for which the wastewater is stored because anaerobic bacteria start degrading dung particles (Møller *et al.* 2002). The authors believe that to manage wastewater from the lairage and paunch section, simple solid–liquid separation equipment such as a hydrasieve, or an externally or internally fed rotary drum filter with suitable screen size may prove better than more sophisticated equipment such as a centrifuge, screw press, or vibratory screen, etc.

Several researchers have also conducted studies to evaluate dissolved air flotation (DAF) performance for treating effluent from a slaughterhouse using both laboratory and full-scale experiments. Ross *et al.* (2000) assessed the DAF unit's performance to treat poultry rendering wastewater with very high SS (43,706 mg/L) and FOG (18,568 mg/L). The DAF was operated with pH adjustment using sulphuric acid and cationic polymer dosing. It was operated at the air to solid (A/S) ratio of 0.0006,

Table 1 | Performances of solid–liquid separation equipment

Solid–liquid separation equipment	Screen Size	Type of slurry	Average flow in m³/h	Influent DM (%)	Effluent DM (%)	% Removal	Separated solids DM (%)	Power consumption[a]	Reference
Screw press	0.25 mm axial wire screen	Swine manure	6.12 4.98	0.84 2.2	0.92 2.43	– –	6.41 34.8	0.239 kWh/m³ 0.625 kWh/m³	Fleming & Macalpine (2003)
Vibratory screen	0.212 mm	Dairy manure	1.56	1.85	1.63	11.89	7.27	0.129 kWh/m³	
Continuous belt microfiltration	Filter fabric opening size: 100 micron	Swine manure	HLR of 0.145 m³/ m²·min	4.7	2.2	53	14–18	Electricity is required for the conveyor belt and air knife for cleaning	Fernandes *et al.* (1988)
				8	3.2	60			
Screw press	0.50 mm screen opening	Dairy farm	6.6	8.32 ± 0.46	5.06 ± 0.35	39.18	24.6 ± 0.67	–	Gooch *et al.* (2005)
	2.25 mm screen opening	Dairy farm	12.75	5.50 ± 0.37	5.19 ± 0.40	0.05	29.3 ± 2.48	–	
	0.75 mm screen opening	Dairy farm	11.21	9.96 ± 0.52	4.93 ± 0.28	50.5	25.3 ± 1.12	–	
Decanter centrifuge	–	Anaerobically digested livestock slurry	4 6 8	5 5 5	2.3 2.3 2.4	54 54 52	21.4 21.6 21.9	Nearly 1 kWh/m³	Giorgia (2013)
Brushed screen separator	1.6 mm screen opening	Cattle slurry	–	60.4 g/kg	33 g/kg	45	129.9 g/kg	0.075 kWh/m³	Gilkinson & Frost (2007)
Decanter centrifuge	Bowl Speed 4,500 rpm	Cattle slurry	–	59.7 g/kg	17.9 g/kg	70	213.4 g/kg	Nearly 1.5 kWh/m³	
Inclined screen	2.5 mm	Dairy farm	–	6,000–7,000 mg/L	2,900 mg/L	51–58	131 g/L	–	Wright (2018)

[a]Power consumption excludes the cost of pumping.

DM, dry matter.

which is nearly 1/10th the minimum value of the A/S ratio reported by Qasim (1998) and Metcalf & Eddy (2003). Both SS and FOG removal efficiency of more than 98% was achieved. Nardi *et al.* (2008) evaluated the performance of DAF with full flow pressurization. SS and FOG removal efficiencies were 43 ± 15% and 49 ± 8%, respectively, when operated with chemical dosing of 24 mg Al^{3+}/L and 1.5 mg anionic polymer. The same DAF unit with full flow pressurization was upgraded to operate with the recycle flow pressurization with a pressure of 4.5 kg/cm^2.

Upgraded DAF under the same chemical dosing resulted in SS and FOG removal efficiencies of 74% and 99%, respectively. Generally, fat removal efficiencies in DAF are more than that of the SS removal efficiencies for two reasons: (1) the nature of fat to travel upward; and (2) its hydrophobic nature (Kitchener & Gochin 1981). Lovett & Travers (1986) measured the performance of a laboratory-scale DAF unit that was treating abattoir wastewater and found that maximum removal efficiencies of SS and FOG were 70% and 95%, respectively, and occurred at influent SS of 1,200–2,105 mg/L and A/S ratios of 0.05–0.06. The study also showed that at low A/S ratios, solid removal occurred by both settling and flotation. On the other hand, at higher A/S ratios, solids removal was mostly due to flotation. Thus SS removal in DAF depends on initial SS concentration, particle size and degree of flocculation, etc. Manjunath *et al.* (2000) evaluated the performance of the laboratory-scale DAF unit to treat wastewater from a slaughterhouse. They found that at a higher A/S ratio of 0.09, SS removal efficiency was nearly 55%, and oils and fat removal efficiency of about 80%. The study also concluded that the DAF unit reduces the strength of wastewater by about 50% and thus fewer loads on the downstream biological treatment unit.

The literature survey also indicated that the current practice is to treat the wastewater from all the slaughtering processes/operations in a combined manner. The most common primary treatment route to treat slaughterhouse wastewater as reported by European Commission (2005), CPCB (2017), IPPC (2003), USEPA (2004), Salminen (2002), USEPA (2002), Gauteng Provincial Government South Africa (2009), EPA Ireland (2008), Enterprise Ireland (2009) and Environment Agency (2009) is presented in Figure 1. It consists of screens, followed by DAF.

The wastewater from the lairage and paunch section exerts a significant SS and COD load (Tritt & Schuchardt 1992; Cumby *et al.* 1999; Mittal 2004). The wastewater from the sticking point has the highest organic strength (Tritt & Schuchardt 1992). Slaughtering operations/processes such as rendering, fleshing, and intestine and tripe washing add FOG and floating solids to the wastewater (Ross *et al.* 2000; Black *et al.* 2013). The FOG presence in wastewater may hinder the performance of screening units due to the accumulation of an oily layer on the sieve openings. The continuous deposition may cause choking of screening equipment after a certain period and necessitate periodic cleaning. At the same time, the screened materials contain dung and meat/flesh/fat pieces that may limit their application in the agricultural fields. The effluent from the screens is usually retained in a holding tank before feeding it into the DAF unit. It was observed that if the effluent is held for a longer period of time, the SS in the effluent tends to travel upwards and forms a thick layer of solids at the top surface of the holding tank.

Moreover, the presence of SS even after screening may offset the performance of DAF units due to an increase in SS load. The wastewater from the lairage and paunch section, which principally contains SS when fed to DAF, also increases the hydraulic load on DAF. Hence, it is necessary to rethink and revisit the primary wastewater treatment in the slaughterhouse industry. The authors feel that the segregation of wastewater from the individual slaughtering process/operation based on the nature of similarity may offer additional benefits such as pollutant specific target treatment, distributed hydraulic load and recovery of useful products for reuse. Therefore, the main objective of this study is to:

- Segregate the wastewater streams from the individual slaughtering process/operations based on the nature of similarity.
- Design, implement and evaluate the performance of an improved primary wastewater treatment system for a slaughterhouse.

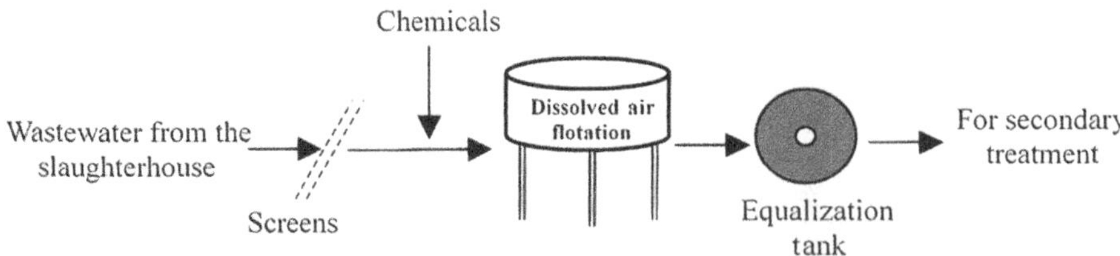

Figure 1 | Typical primary wastewater treatment route in a slaughterhouse.

The improved primary treatment system in this study is the result of a detailed literature review, physico-chemical characterization of wastewater emanating from the individual slaughterhouse processes/operation, water consumption patterns and close observations.

MATERIALS AND METHODS

Study area and approach

This study was carried out in a slaughterhouse with a slaughtering capacity of 1,000 buffaloes per day and is located in Uttar Pradesh, India. The slaughterhouse follows International Standards and is Hazard Analysis Critical Control Point, ISO 22000:2005, and ISO 9001:2015 certified. The starting point for developing an improved primary treatment was the implementation of segregation of streams; hence a detailed plan was worked out to segregate the various wastewater streams depending on their similar nature, and taking into account the existing wastewater conveyance network, topography and ease of execution. Figure 2 presents the implementation plan for the segregation of streams in the slaughterhouse industry. Dung containing streams from lairage and paunch room were segregated, and FOG containing streams from fleshing, intestine and tripe washing and rendering were separated to facilitate specific treatment and removal of targeted pollutants. Similarly, the bloodstream from the slaughtering area and salt-containing stream from the hide storage area were also separated to prevent its entry into the ETP. This article specifically deals with the wastewater from the lairage and paunch section, rendering, intestine and tripe washing, fleshing and the carcass hall for treatment, removal and recovery of SS and FOG.

The improved primary wastewater treatment system

The wastewater from the lairage section was treated independently with inclined 500 μm parabolic screens referred to as a hydrasieve. The filtrate from the hydrasieve was then mixed with the paunch room wastewater, and the resulting mixture of streams was treated with a 200 μm externally fed rotary drum filter (EFRDF). The filtrate from the EFRDF was then taken to the primary clarifier after adding alum as a coagulant to remove SS completely. The effluents from the lairage and paunch section were kept separate because the dung particles from the lairage section are completely digested and were used by local farmers as manure in their fields, whereas the dung particles from the paunch section are partially digested.

Using separate screening equipment in the form of 500 μm hydrasieve and 200 μm drum filter for lairage and paunch sections, respectively, substantially reduced SS loads, improved the performance of the treatment units and increased the overall life of the equipment. The FOG containing streams from rendering, fleshing, and intestine and tripe washing contains floating solids. Thus, the FOG containing streams were first routed through the skimming tank, wherein floating solids are recovered from the top.

The effluent from the skimming tank was then fed to the DAF unit. It is important to mention here that using a skimming tank before the DAF removes a substantial fraction of FOG and prevents the floating solids from forming a thick layer at the top surface of the feed/holding tank. This top layer subsequently starts degrading, creates unaesthetic conditions due to foul odour, and it becomes difficult to remove the thick layer of floating solids periodically. Moreover, removing floating solids in a skimming tank reduces the significant load of floating solids, which may upset DAF performance. The workings of the hydrasieve, EFRDF, skimming tank and DAF is explained in the following sections.

Hydrasieve

A hydrasieve is a curved concave type of stationary screen. Wastewater is pumped to the overflow trough located at the top of the hydrasieve screen. As the wastewater slides on the screen's surface, the solid materials are trapped in the screen. Separated particles move down to the bottom and fall in a hopper due to the effect of subsequent flow impulsive force, the angular orientation of the screen and the weight of solids. These screens can be manufactured either in perforations or in a mesh pattern. The screen wires, referred to as wedge wire, are triangular in cross-section, which helps liquids to attach hydraulically to the bars, leaving any solids on the upward portion of the screen plate (Ross *et al.* 1980; USEPA 2002). The hydrasieve offers advantages including low power requirement, ease of operation, minimum moving parts and there is no operator contact with the liquid during the cleaning operation.

Externally fed rotary drum filter

An EFRDF has a screening or straining medium which is mounted on a cylinder that rotates at a fixed rotation per minute (rpm). The wastewater flows into the top of the unit and passes through the interior, with solids collected on the exterior.

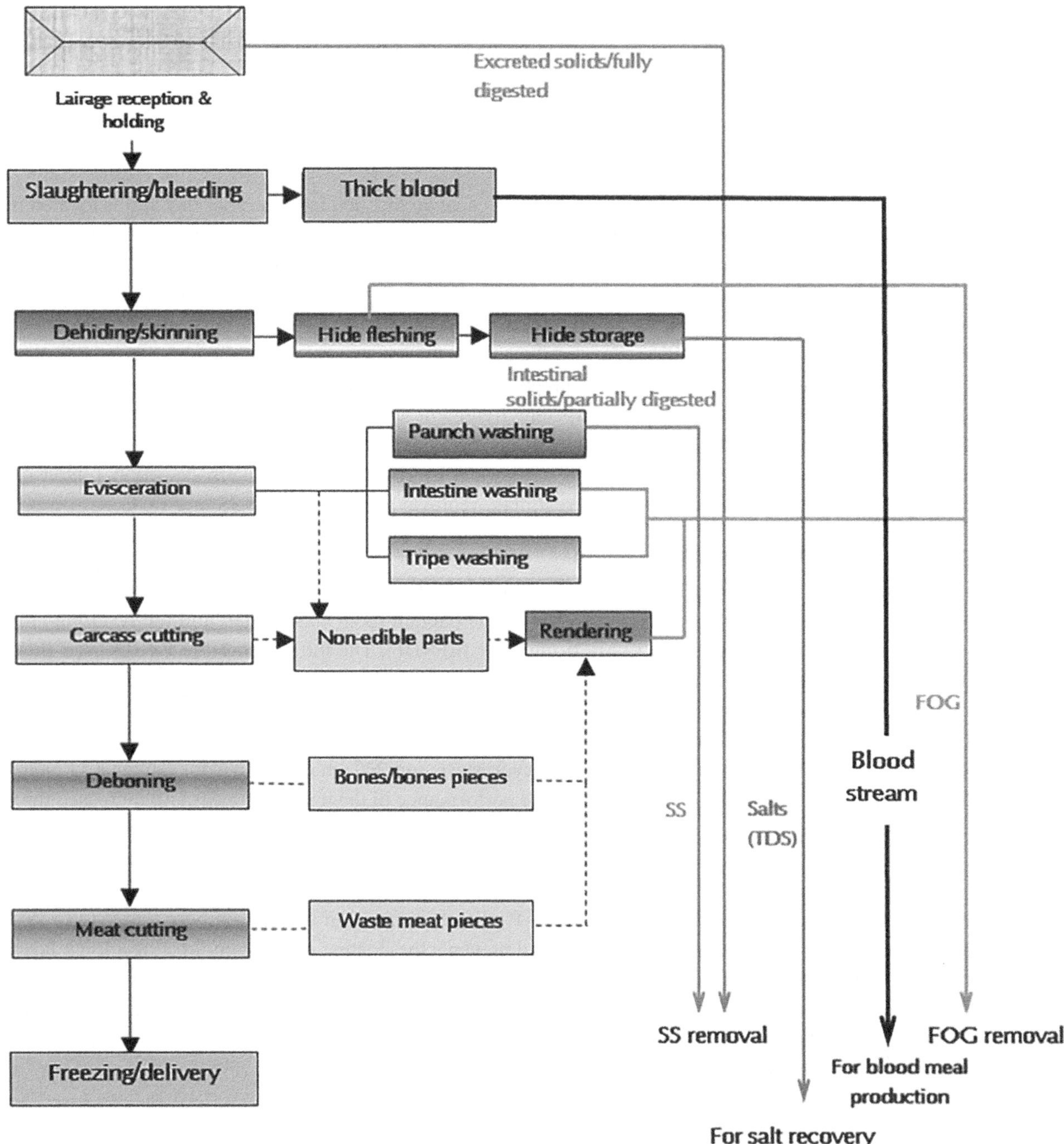

Figure 2 | Various slaughtering processes/operations and their segregation.

Drum screens are mostly fabricated either in mesh or wedge wire only, and perforations drum screens are quite uncommon. To prevent clogging, the screen is sprayed continuously with water using nozzles. These types of screens are least susceptible to clogging caused by solids (Ross *et al.* 1980; USEPA 2002).

Skimming tank

Skimming tanks are generally installed as a pre-treatment unit to remove floating substances like oils, fats, waxes, free fatty acids, soaps and floating debris (Rao 2005). A skimming tank is a chamber arranged such that floating matter rises and remains on the surface of wastewater until removed manually or mechanically. The effluent free of FOG is collected continuously through an outlet located at a certain depth with curtain walls and a deep scum board (Punmia & Jain 2005). Floating matters are removed either manually or with the help of mechanical equipment.

Dissolved air flotation

Flotation is a very effective method of separating solids and liquid to remove low-density particles that tend to float. Air flotation is used to thicken the solids, and separation of solids is achieved by introducing fine air bubbles. The DAF has four major components: compressor, high-pressure pump, saturator and flotation chamber. The pressurized air is brought in contact with the recycled wastewater in a saturator, and sufficient time is given to allow the air to dissolve in wastewater. The mixture of air and wastewater is introduced in the main chamber. Due to the sudden release in pressure, microbubbles are formed that attach themselves to particles and travel upwards to form a floating layer (Srinivasan & Viraraghavan 2009).

After segregating the streams based on their nature of similarity and implementing improved primary treatment systems as described above, individual streams were studied for the quantitative and qualitative assessment, which are addressed later.

Secondary biological treatment system

Substantial removal of SS and FOG was achieved after successful implementation of an improved primary treatment system, which ensured the reduction in organic load on downstream secondary treatment system. Accordingly, the effluent was further treated a UASB reactor followed by an activated sludge process (ASP). The performance of the secondary treatment system was assessed by collecting integrated samples for 10 h at the inlet and outlet of the UASB reactor and ASP–secondary clarifier system. All the samples were analyzed for SS, COD, BOD, total Kjeldahl nitrogen (TKN) and total phosphate (TP) in compliance with the procedures in the Standard Methods for the examination of water and wastewater, American Public Health Association (APHA), 2005. The results presented on the performance evaluation of the secondary biological treatment system are representative of three sampling events after the successful implementation and commissioning of the improved primary treatment system. Other operating parameters such as food to microorganism ratio, hydraulic retention time (HRT), solids retention time (SRT), mixed liquor suspended solids (MLSS), oxygen uptake rate, specific oxygen uptake rate and sludge volume index for ASP were calculated according to Metcalf & Eddy (2003) and Qasim (1998).

RESULTS AND DISCUSSION

An improved primary wastewater treatment system was designed and installed on a full scale. The improved primary treatment system was studied for a period of one year for its effectiveness, operational difficulties, ease and efficacy. The quantitative and qualitative assessments of SS and FOG streams after segregation are shown in Table 2. Figure 3 shows pictures of segregated blood, dung and FOG streams from various slaughtering operations.

The segregated bloodstream (86 L/buffalo, COD 31,600–121,600 mg/L) was sent to the decanter centrifuge for blood meal production, thereby preventing the significant organic load from entering the ETP. The average daily quantity of blood meal produced was 1.5 tonnes per day (T/day) with 90% protein and 4% moisture content. The salt stream from the hide storage section (~5 L/buffalo, TDS 25–32%) was also separated for salt recovery. There is the potential to recover salts to the tune of 2.5–3 T/day. Intestinal contents, which are partially digested, are also a major pollutant in slaughterhouses (Carawan & Pilkington 1986). The wet weight of a paunch material ranges between 22 and 31 kg per cattle (Carawan *et al.* 1979). The paunch is cut with the help of a knife, and wet paunch material is taken out manually, and any excess solids stuck on the inner wall of the paunch are rinsed with water, which generates wastewater. In a slaughterhouse, the average water consumption per buffalo was 1,114 L, and the corresponding wastewater generation was 916–1,089 L (Shende *et al.* 2021).

Table 2 | Details of effluent generation, concentrations of major pollutants after segregation (capacity 1,000 buffaloes/day)

Stream(s)	Flow (m³/d)	Concentration	Target parameters
Lairage	75–90	SS: 0.6–1.5%	SS, COD
Paunch	200–250	SS: 1.0–2.5%	SS, COD
Hide fleshing, intestine and tripe washing, rendering	250–300	FOG: 1.2–3.5%	FOG
Blood	80–90	COD: 31,600–121,600 mg/L	COD, BOD
Salts	8–10	TDS: 20.0–22.0%	TDS

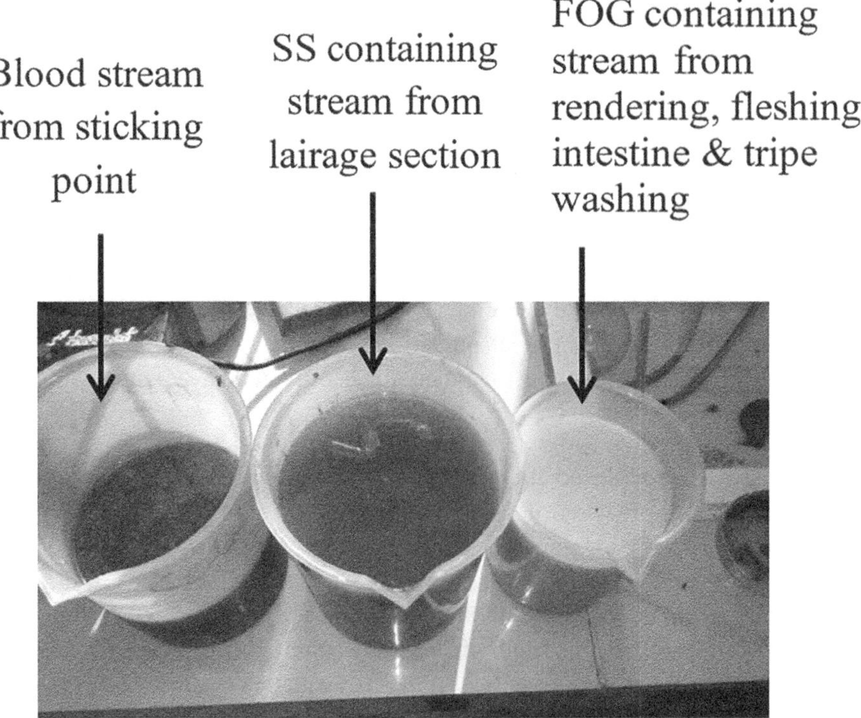

Figure 3 | Photograph of wastewater streams after segregation in a slaughterhouse.

Efficacy of an improved primary wastewater treatment system

The schematic and performance of an improved primary treatment system are shown in Figure 4 and the results presented are the average of five sampling events taken during 2019. The samples were taken at the end of each hour during the operation of the individual solid–liquid separation equipment and mixed to form the representative sample for analysis.

The average wastewater generation from the lairage section varies between 75 and 90 m^3/d, and the SS removal efficiency from the hydrasieve was around 75% at an HLR of 6.25 m^3/m^2/h. Usually, HLR for hydrasieves varies between 2.4 and 72 m^3/m^2/h (Metcalf & Eddy 2003). The SS removal efficiency using a hydrasieve in this study was higher than that reported by Wright (2018). This may be attributed to the opening size used for the hydrasieve screens. Wright (2018) used a hydrasieve with an opening of 2.5 mm rather than 500 μm, as in this study. It is important to maintain harmony between screen size and HLR because the reduction in screen opening reduces applicable HLR. The same was also observed in this study, where an increase in HLR beyond 6.25 m^3/m^2/h resulted in the splashing of water over the screen surface and a reduction in SS removal efficiency.

Nevertheless, the hydrasieve with a screen opening of 500 μm worked satisfactorily at an HLR of 6.25 m^3/m^2/h with steady SS removal efficiency. The wastewater from the paunch section, carcass hall, meat packaging and refrigeration was mixed with the filtrate obtained from the hydrasieve. The resultant mixture had a flow rate in the range of 525–610 m^3/d and was then fed to the drum filter at an HLR of 10 m^3/m^2/h, which achieved 55% average SS removal efficiency. The outlet of the drum filter had 1,544 $\pm$ 330 mg/L of SS and 1,804 $\pm$ 168 mg/L of COD concentrations. It was observed that the SS at the outlet of the drum filter was mostly colloidal. These colloidal solids were removed using coagulation with an alum dose of 250–300 mg/L. After coagulation and subsequent settling in the primary clarifier, SS and COD removal efficiencies of nearly 86% and 52% were achieved, respectively. The results obtained in this study were comparable with the studies done by Núñez *et al.* (1999) and Aguilar *et al.* (2005) to treat slaughterhouse wastewater using coagulation with alum. The sludge volume in this study was found to be 150 $\pm$ 30 mL/L, which was higher than the reported value of 85 mL/L while treating slaughterhouse wastewater by Satyanarayan *et al.* (2005). This is because the initial SS concentration in the study carried out by Satyanarayan *et al.* (2005) was only 310 mg/L rather than 1,560 $\pm$ 350 mg/L in the present study. The effluent obtained at the outlet of the primary clarifier was found to have very low organic strength with COD and BOD concentrations of 850 $\pm$ 125 and 596 $\pm$ 90 mg/L, respectively. The sludge from the primary clarifier was dewatered using a filter press.

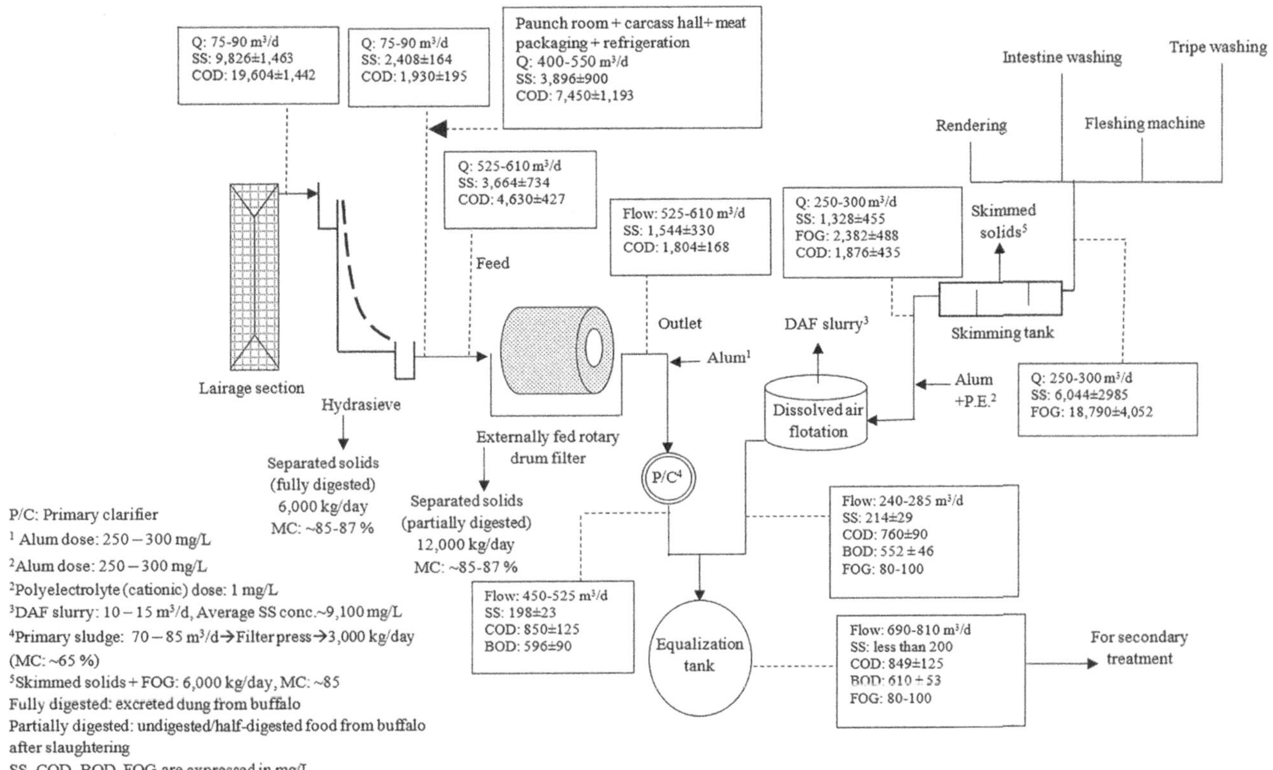

Figure 4 | Schematic diagram of Improved Primary Treatment System for Slaughterhouse Wastewater and its performance.

The average moisture content of the dewatered sludge was 65% and the remaining dry solids (DS) was 35%. The dewatered solids from the filter press are around 3,000 kg/day.

The separated solids from the hydrasieve, EFRDF and filter press were 6,000, 12,000, and 3,000 kg/day, respectively. The average moisture content (MC) of the separated solids from the hydrasieve and EFRDF was 85%, and for the filter press was 65%. The separated solids were dewatered, dried and manufactured into briquettes with a gross calorific value of 3,032 Kcal/kg.

The skimming tank was efficient in removing FOG, meat/flesh and fat pieces. The SS concentrations at the inlet of the skimming tank had a wide variation with concentrations of 6,044 ± 2,985 mg/L. However, the SS concentration at the outlet of the skimming tank outlet was relatively low at 1,328 ± 455 mg/L, resulting in nearly 75% removal efficiency. As far as the removal of FOG is concerned, the skimming tank achieved more than 90% removal efficiency. The outlet of the skimming tank was then fed to the DAF unit.

The A/S ratio in the DAF in this study was more than the values reported by Ross *et al.* (2000) and Nardi *et al.* (2008). This is mainly because of the air injection rate and the recycle pressure, which was maintained at 3–3.5 m³/h and 6.5 kg/cm², respectively. The SS removal efficiency in the present study is comparable to the values reported by Nardi *et al.* (2008). However, the FOG removal efficiency in the present study was comparatively lower than the values reported by Ross *et al.* (2000) and Nardi *et al.* (2008). After the DAF unit, the effluent obtained had FOG concentrations in the range of 80–100 mg/L. The operating parameters and the comparative assessment of the performance of the DAF unit are shown in Table 3.

The effluents obtained from the DAF unit and primary clarifier were mixed in the equalization tank. The resultant wastewater was found to have COD and BOD concentrations of 849 ± 125 and 610 ± 53 mg/L, and SS and FOG concentrations of less than 200 and 100 mg/L, respectively.

The summary of SS removed through the multi-intervention approach consisting of the hydrasieve, EFRDF, skimming tank and DAF along with the quantities of valuable resources from blood and salt stream is presented in Table 4. It is evident from Table 4 that the hydrasieve and EFRDF separated significant SS from the effluent. Moreover, a simple intervention like a skimming tank proved very efficient and removed FOG significantly (4–4.5 kg/buffalo).

Table 3 | Operating parameters and comparative assessment of the performance of DAF unit

Operating parameters	FOG streams from fleshing, rendering, and intestine and tripe washing at the outlet of the skimming tank (present study)	Wastewater from poultry rendering facility (Ross *et al.* 2000)	Wastewater from poultry slaughterhouse (Nardi *et al.* 2008)
Total flow (m^3/h)	68	74.95	105.43
Recycle rate (m^3/h)	22.44	20.44	42.18
Percentage recycling ratio (%)	33	27.27	40
Recycle pressure (kg/cm^2)	6.5	5.6	4.5
Air injection rate (m^3/h)	3–3.5	1.84	–
Surface area (m^2)	11.93	16.72	32
HLR (feed only) (m^3/m^2/h)	3.82	4.48	2.4 ± 0.2
HLR (including recycle) (m^3/m^2/h)	5.69	5.70	4.61
Solid loading rate (kg/m^2/h)	4.98–10.16	195.69	0.916 ± 0.20
Air to solid ratio (mL/mg)	0.016	0.0006	0.030 ± 0.008
Influent SS mg/L	$1,328 \pm 455$	43,706	861 ± 204
Influent FOG mg/L	$2,382 \pm 488$	18,568	182 ± 29
Effluent SS mg/L	214 ± 29	262	224 ± 53
Effluent FOG mg/L	80–100	72	<2

Table 4 | Summary of solids and FOG recovered in treatment units

Treatment unit	SS separated/DS/FOG removed
Hydrasieve	SS: 6 kg/buffalo (digested, MC: ~85–87%), DS: 0.9–1 kg/buffalo
EFRDF	SS: 12 kg/buffalo (partially digested, MC: ~85–87%), DS: 1.8–2 kg/buffalo
Primary clarifier → filter press	SS: 2–3 kg/buffalo (MC: ~60–65), DS: 0.8–1 kg/buffalo
SS recovered from lairage and a paunch section either by dry scrapping or by manually emptying the paunch	SS: 10–13 kg/buffalo (MC: ~85%), DS: 1.5–2 kg/buffalo
Skimming tank	SS: 1.1–1.4 kg/buffalo FOG: 4–4.5 kg/buffalo
DAF	SS: ~0.3 kg/buffalo FOG: 0.5–0.6 kg/buffalo
Separated bloodstream	COD: 31,600–121,600 mg/L, daily blood meal production 1.5 T/day with protein content of ~90%
Separated salt stream	Potential salt recovery: ~ 2,000–2,200 kg/day

Secondary biological treatment

The secondary biological treatment system and its performance is presented in Figure 5. During the study period, it was found that the concentrations of SS, COD, BOD, TKN and TP of the primary treated wastewater were 187 ± 20, 823 ± 186, 540 ± 104, 66 ± 18, and 5.3 ± 0.69 mg/L, respectively. The same wastewater was fed to full-scale UASB reactors at an average organic loading rate (OLR) of 1.54 ± 0.34 kg COD·m^3/d with an HRT of 12.8 h at an upflow velocity V_{up} of 0.15 m/h. The average COD and BOD removal efficiencies after the UASB were 33% and 37%, respectively, which are less than the

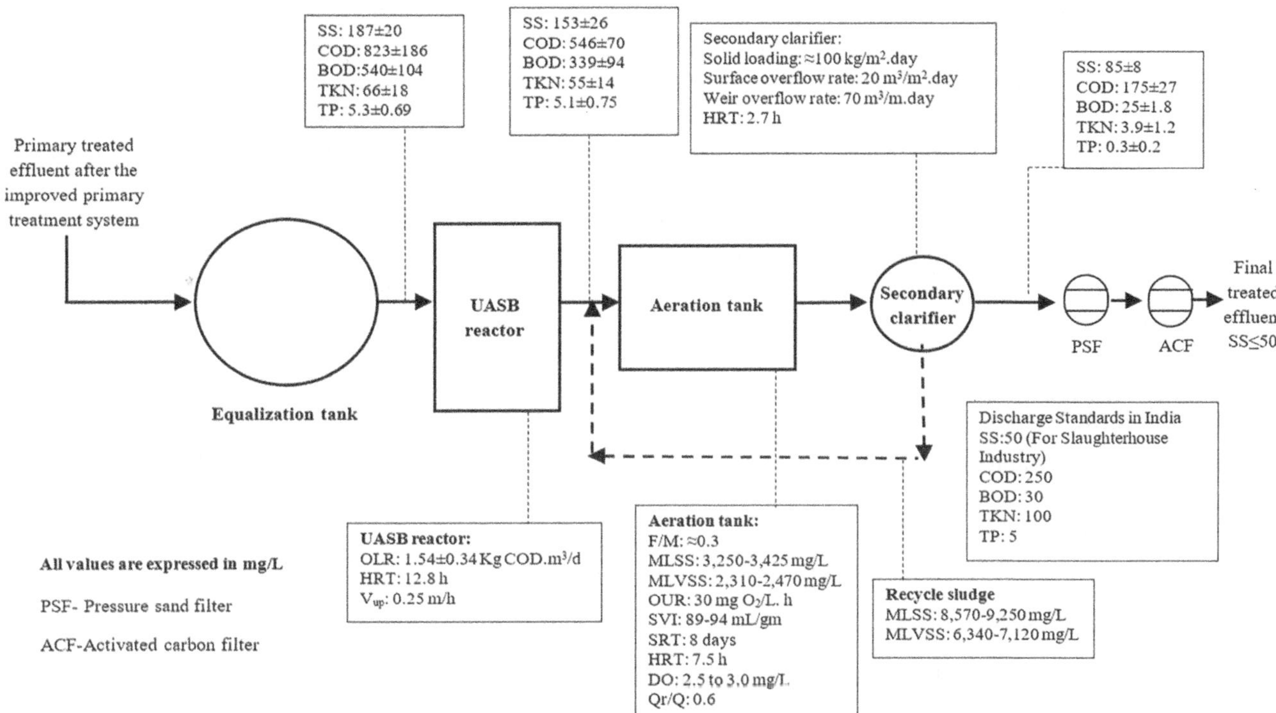

Figure 5 | Performance evaluation of secondary biological treatment system.

reported values by Sayed *et al.* (1987). They operated the laboratory-scale UASB at OLR 2.5–4.0 kg COD·m³/d with an HRT of 9 h and achieved 56% COD removal efficiency while treating slaughterhouse wastewater with an initial COD concentration of 1,086 mg/L. The difference in the results can be attributed to the huge difference in the scale of operation. In the present case, UASB was able to reduce the organic load to aerobic biological treatment systems by at least 35%. No biomass washout was observed during the reactor operation. The outlet from the UASB was fed to the ASP, which was operated at an SRT of 8 days and HRT of 7.5 h. The recycle ratio of return sludge to the influent was maintained at 0.6. The ASP performed satisfactorily with the average COD removal efficiency of 68%. The average BOD removal efficiency of the ASP was 92%, proving that the ASP was effective in bringing down the BOD of treated effluent to meet the effluent discharge norms to ≤30 mg/L. The MLSS to mixed liquor volatile suspended solids ratio was 0.7, and the oxygen uptake rate was 30 mg O₂/L·h indicating the presence of healthy biomass. The sludge volume index was found to be between 89 and 94 mL/g showing good settling of MLSS. This is also evident from the solids concentration of the return activated sludge, which varied between 8,570 and 9,250 mg/L. The TKN and TP values after ASP were also found to be meeting the effluent discharge standards.

As per the revised effluent discharge standards for slaughterhouse industries (Environmental Protection Rules, 1986), SS needs to be brought down to less than 50 mg/L. However, the SS concentration in the ASP outlet was 85 ± 8 mg/L, which is higher than the regulatory standards. However, the slaughterhouse industry operates a tertiary treatment system consisting of pressure sand filter followed by activated carbon filter and/or polishing ponds which can take care of the SS. The combined performance of the UASB followed by the ASP was satisfactory, and the treated effluent at the outlet of the ASP met the effluent discharge standards in terms of COD, BOD, TKN and TP. The SS concentration was further reduced with the help of a tertiary treatment system consisting of a pressure sand filter and an activated carbon filter.

CONCLUSIONS

One of the major challenges in dealing with slaughterhouse industrial wastewater is the effective primary treatment system for removing SS and FOG, which is a major hurdle in downstream treatment. Conventionally, wastewater generated from various slaughtering operations is treated in a combined manner. The segregation of streams based on the nature of similarity and target-specific pollutant treatment was proven to be very effective in managing slaughterhouse wastewater. Using simple solid–liquid separation equipment such as a hydrasieve (500 μm) and EFRDF (200 μm) with suitable screen sizes proved was effective compared to sophisticated equipment such as centrifuge, screw press, vibratory screens, etc. The intervention of the skimming tank before DAF was found to reduce considerable SS and FOG load on the DAF. The effluent obtained After using the improved primary wastewater treatment system, the effluent obtained had SS and FOG less than 200 mg/L and 100 mg/L, respectively. The primary treated effluent was managed with a UASB reactor followed by a conventional ASP. The secondary biological treatment system performed smoothly without experiencing operational difficulties and achieved effluent discharge standards in terms of COD, BOD, TKN and TP. Segregation of similar streams and making provisions for removing dung solids SS and FOG through multi-intervention approach offered benefits in relation to resource recovery, distributed hydraulic load reduced waste load, and ease in wastewater treatment. Thus, the segregation of streams from the individual slaughterhouse processes/operation based on its nature of similarity and separate treatment is key to effectively managing wastewater in the slaughterhouse industry.

ACKNOWLEDGEMENTS

The help by M/s FEPL, Rampur (Uttar Pradesh), and Barabanki (Uttar Pradesh) is gratefully acknowledged.

CONFLICT OF INTEREST

The authors declare no conflict of interest.

DATA AVAILABILITY STATEMENT

All relevant data are included in the paper or its Supplementary Information.

REFERENCES

Aguilar, M. I., Sáez, J., Lloréns, M., Soler, A., Ortuño, J. F., Meseguer, V. & Fuentes, A. 2005 Improvement of coagulation-flocculation process using anionic polyacrylamide as coagulant aid. *Chemosphere* **58**, 47–56. https://doi.org/10.1016/j.chemosphere.2004.09.008.
American Public Health Association (APHA) 2005 *Standard Methods for the Examination of Water and Wastewater*, 21st edn. APHA, Washington, DC.
Black, M., Canova, M., Rydin, S., Scalet, B. M., Roudier, S. & Sancho, L. D. 2013 Best available techniques (BAT) reference document on for the tanning of hides and skin. *European Commission*. http://dx.doi.org/10.2788/13548.
Carawan, R. E. & Pilkington, D. H. 1986 *Reduction in Waste Load from a Meat Processing Plant-Beef*. North Carolina Agricultural extension service, Raleigh, North Carolina.
Carawan, R. E., Chambers, J. V. & Zall, R. R. 1979 *Meat Processing Water and Wastewater Management*. North Carolina Agricultural Extension Service, Raleigh, North Carolina.
Central Pollution Control Board (CPCB) 2017 *Revised Comprehensive Industry Document on Slaughterhouses*. Government of India.
Cumby, T. R., Brewer, A. J. & Dimmock, S. J. 1999 Dirty water from dairy farms, I: biochemical characteristics. *Bioresource Technology* **67**, 155–160. https://doi.org/10.1016/S0960-8524(98)00104-7.
Department of Agriculture and Rural Development 2009 *Guideline Manual for the Management of Abattoirs and Other Waste of Animal Origin*. Gauteng Provincial Government, South Africa.
Enterprise Ireland 2009 *Sustainable Practices in Irish Beef Processing*. Enterprise Ireland, Dublin.
Environment Agency 2009 *How to Comply with Your Environmental Permit. Additional Guidance for The Red Meat Processing (Cattle, Sheep and Pigs) Sector (EPR 6.12)*. Environment Agency, Bristol.
Environmental Protection Agency 2008 *BAT Guidance Note on Best Available Techniques for the Slaughtering Sector*, 1st edn. Environmental Protection Agency, Wexford, Ireland.
Environmental Protection Rules, Government of India 1986 Standards for discharge of effluents from slaughterhouse, meat processing units and seafood industry. Notified on 28.10.2016.
European Integrated Pollution Prevention and Control (IPPC) Bureau 2003 *Reference Document on Best Available Techniques in Common Waste Water and Waste Gas Treatment/Management Systems in the Chemical Sector, 472*. Available from: http://eippcb.jrc.ec.europa.eu/reference/cww.html.
European Commission 2005 *Reference Document on Best Available Techniques in the Slaughterhouses and Animal by-Products Industries*. Brussels.
Fernandes, L., Mckyes, E. & Obidniak, L. 1988 Performance of a continuous belt microscreening unit for solid liquid separation of swine waste. 151–155.
Fleming, R. & Macalpine, M. 2003 *Evaluation of Mechanical Liquid/Solid Manure Separators*. Prepared for Ontario Pork, Project No. 02/39. Ridgetown College and University of Guelph, Ridgetown and Ontario.
Ghaffour, N. 2004 Modeling of fouling phenomena in cross-flow ultrafiltration of suspensions containing suspended solids and oil droplets. *Desalination* **167**, 281–291. https://doi.org/10.1016/j.desal.2004.06.137.

Gilkinson, S. & Frost, P. 2007 *Evaluation of Mechanical Separation of Pig and Cattle Slurries by a Decanting Centrifuge and a Brushed Screen Separator*. Agri-Food and Biosciences Institute, Hillsborough, Northern Ireland, pp. 1–47.

Giorgia, C. 2013 PhD. Thesis: *Assessment of Different Solid-Liquid Separation Techniques for Livestock Slurry*. University DEGLI Studi DI Milano.

Gooch, C. A., Inglis, S. F. & Czymmek, K. J. 2005 Mechanical solid-liquid manure separation: Performance evaluation on four New York State dairy farms – a preliminary report. In *ASAE Annual International Meeting*. p. 0300. https://doi.org/10.13031/2013.19506

Kitchener, J. A. & Gochin, R. J. 1981 The mechanism of dissolved air flotation for potable water: basic analysis and a proposal. *Water Research* **15**, 585–590. https://doi.org/10.1016/0043-1354(81)90021-X.

Lovett, D. A. & Travers, S. M. 1986 Dissolved air flotation for abattoir wastewater. *Water Research* **20**, 421–426. https://doi.org/10.1016/0043-1354(86)90188-0.

Manjunath, N. T., Mehrotra, I. & Mathur, R. P. 2000 Treatment of wastewater from slaughterhouse by DAF-UASB system. *Water Research* **34**, 1930–1936. https://doi.org/10.1016/S0043-1354(99)00337-1.

Metcalf & Eddy 2003 *Wastewater Engineering*, 4th edn. Tata McGraw-Hill Publishing Company Limited, New Delhi, India.

Miranda, L. A. S., Henriques, J. A. P. & Monteggia, L. O. A. 2005 Full-scale UASB reactor for treatment of pig and cattle slaughterhouse wastewater with a high oil and grease content. *Brazilian Journal of Chemical Engineering* **22**, 601–610. https://doi.org/10.1590/S0104-66322005000400013.

Mittal, G. S. 2004 Characterization of the effluent wastewater from abattoirs for land application. *Food Reviews International* **20**, 229–256. https://doi.org/10.1081/FRI-200029422.

Møller, H. B., Sommer, S. G. & Ahring, B. K. 2002 Separation efficiency and particle size distribution in relation to manure type and storage conditions. *Bioresource Technology* **85**, 189–196. https://doi.org/10.1016/S0960-8524(02)00047-0.

Nardi, I. R., Fuzi, T. P. & Nery, V. D. 2008 Performance evaluation and operating strategies of dissolved-air flotation system treating poultry slaughterhouse wastewater. *Resource Conservation and Recycling* **52**, 533–544. https://doi.org/10.1016/j.resconrec.2007.06.005.

Núñez, L. A., Fuente, E., Martínez, B. & García, P. A. 1999 Slaughterhouse wastewater treatment using ferric and aluminium salts and organic polyelectrolites. *Journal of Environmental Science & Health – Part A* **34**, 721–736. https://doi.org/10.1080/10934529909376861.

Punmia, B. C. & Jain, A. K. 2005 *Wastewater Engineering*. Laxmi Publications Private Limited, New Delhi, India.

Qasim, S. R. 1998 Wastewater Treatment Plants – Planning, Design and Operation, second ed. Technomic Publishing Company, Lancaster, Pennsylvania.

Rao, P. V. 2005 *Textbook of Environmental Engineering*. Prentice-Hall of India Private Limited, New Delhi.

Ross, S. A., Guo, P. H. M. & Jank, B. E. 1980 *Design and Selection of Small Wastewater Treatment System. Environment Protection Service Report Series*. Environment Canada Ottawa, Canada.

Ross, C. C., Smith, B. M. & Valentine, G. E. 2000 Rethinking Dissolved Air Flotation (DAF) design for industrial pre-treatment. In *WEF and Purdue University Industrial Wastes Technical Conference*, St Louis, Missouri, 21–24 May.

Salehion, A. R., Minaei, S. & Razavi, S. J. 2013 Design and performance evaluation of a screw press separator for separating dairy cattle manure. *International Journal of Agronomy and Plant Production* **4**, 3849–3858.

Salminen, E. 2002 *Finnish Expert Report on Best Available Techniques in Slaughterhouses and Installations for the Disposal or Recycling of Animal Carcasses and Animal Waste. The Finnish Environment, 539*. Finnish Environment Institute, Helsinki, Finland.

Satyanarayan, S., Ramakant & Vanerkar, P. 2005 Conventional approach for abattoir wastewater treatment. *Environmental Technology* **26**, 441–447. https://doi.org/10.1080/09593332608618554.

Sayed, S. K. I., Van Campen, L. & Lettinga, G. 1987 Anaerobic treatment of slaughterhouse waste using a granular sludge UASB reactor. *Biological Wastes* **21** (1), 11–28.

Shende, A. D., Dhenkula, S., Waghambare, A., Rao, N. N. & Pophali, G. R. 2021 Water consumption, wastewater generation and characterization of a slaughterhouse for resource conservation and recovery. *Water Practise & Technology*. https://doi.org/10.2166/wpt.2021.122.

Srinivasan, A. & Viraraghavan, T. 2009 Dissolved air flotation in industrial wastewater treatment. In: *Water and Wastewater Treatment*, (edited by Vigneswaran S.). Encyclopedia of Life Support Systems (EOLSS) Publisher, Oxford, UK.

Sunder, G. C. & Satyanarayan, S. 2013 Efficient treatment of slaughter house wastewater by anaerobic hybrid reactor packed with special floating media. *International Journal of Chemical and Physical Science* **2**, 73–81.

Tritt, W. P. & Schuchardt, F. 1992 Materials flow and possibilities of treating liquid and solid wastes from slaughterhouses in Germany. A review. *Bioresource Technology* **41**, 235–245. https://doi.org/10.1016/0960-8524(92)90008-L.

United States Environmental Protection Agency (USEPA) 2002 *Development Document for the Proposed Effluent Limitations Guidelines and Standards for the Meat and Poultry Products Industry Point Source Category*. Washington DC.

United States Environmental Protection Agency (USEPA) 2004 *Effluent Limitations Guidelines and New Source Performance Standards for the Meat and Poultry Products Point Source Category, 69, (173)*. Washington DC.

Wright, W. F. 2018 *Dairy Manure Particle Size Distribution, Properties, and Implications for Manure Handling and Treatment. The Society for Engineering in Agricultural, Food and Biological System, Paper No. 054105*. https://doi.org/10.13031/2013.19507.

First received 21 July 2021; accepted in revised form 21 January 2022. Available online 3 February 2022

doi: 10.2166/wst.2021.364

Fabricating novel PVDF-g-IBMA copolymer hydrophilic ultrafiltration membrane for treating papermaking wastewater with good antifouling property

Yujia Tong, Wenlong Ding, Lijian Shi and Weixing Li*

State Key Laboratory of Materials-Oriented Chemical Engineering, College of Chemical Engineering, Nanjing Tech University, Nanjing 210009, China
*Corresponding author. E-mail: wxli@njtech.edu.cn

ABSTRACT

Ultrafiltration membranes are widely used for the treatment of papermaking wastewater. The antifouling performance of polyvinylidene fluoride (PVDF) ultrafiltration membranes can be improved by changing the hydrophilicity. Here, a novel amphiphilic copolymer material, PVDF grafted with N-isobutoxy methacrylamide (PVDF-g-IBMA), was prepared using ultraviolet-induced Cu(II)-mediated reversible deactivation radical polymerization. The amphipathic copolymer was used to prepare ultrafiltration membrane via NIPS. The prepared PVDF-g-IBMA ultrafiltration membrane was estimated using ^{1}H NMR, FT-IR, and DSC. The contact angle, casting viscosity, and the permeation performance of the PVDF-g-IBMA ultrafiltration membrane were also determined. The pure water flux, bovine serum albumin removal rate, and pure water flux recovery rate of the PVDF-g-IBMA ultrafiltration membrane were 432.8 L·m^{-2}·h^{-1}, 88.4%, and 90.8%, respectively. Furthermore, for the treatment of actual papermaking wastewater, the chemical oxygen demand and turbidity removal rates of the membrane were 61.5% and 92.8%, respectively. The PVDF-g-IBMA amphiphilic copolymer ultrafiltration membrane exhibited good hydrophilicity and antifouling properties, indicating its potential for treating papermaking wastewater.

Key words: antifouling, papermaking wastewater, ultrafiltration membrane, UV-induced Cu(II)-mediated RDRP

HIGHLIGHTS

- A novel amphiphilic copolymer PVDF-g-IBMA prepared by UV-induced RDRP.
- Good hydrophilic and anti-fouling properties of PVDF-g-IBMA copolymer UF membrane.
- The MWCO, BSA rejection rate and pure water recovery ratio were 43.7 kDa, 88.4% and 90.8%, respectively.
- For treating actual papermaking wastewater, the COD removal rate reached 61.5%.

NOMENCLATURE

J	Membrane permeation flux, L·m^{-2}·h^{-1}
V	Volume of pure water permeated through the membrane, L
A	Effective surface area of the membrane, m^2
Δt	Permeation time, h
R	BSA rejection rate, %
C_p	Protein concentrations in the permeate, g·L^{-1}
C_f	Protein concentrations in the feed, g·L^{-1}
FRR	Pure water recovery ratio, %
J_r	Original pure water flux, L·m^{-2}·h^{-1}
J_w	Recovered water flux, L·m^{-2}·h^{-1}
COD	Chemical oxygen demand, mg·L^{-1}

1. INTRODUCTION

Water shortage is a major crisis affecting human health and sustainable economic and social development. Ultrafiltration membrane (UF) technology is used for the treatment of papermaking wastewater to comprehensively remove impurities from wastewater (Gönder *et al.* 2012; Toczyłowska-Mamińska 2017; Haq *et al.* 2020). Particularly, high-performance UF membranes are important for water purification. The materials used for preparing UF membranes determine the efficiency of separation technology (Shen *et al.* 2020; Sun *et al.* 2020; Zhang *et al.* 2020). Commercial polyvinylidene fluoride (PVDF) membranes exhibit a low surface energy and high hydrophobicity, which facilitates the adsorption of proteins or organics on the pores and surface of membranes (Gao *et al.* 2011; Gao *et al.* 2019; Chen *et al.* 2021). Then, the significantly low permeation flux of fouling PVDF membranes owing to their high filtration resistance has restricted their further application . In addition, the issues associated with cleaning PVDF membranes affect their application cost and reduce their service life (Sri Abirami Saraswathi *et al.* 2018; Sierke & Ellis 2019). Thus, it is important to improve the hydrophilic properties of PVDF UF membranes to extend their application.

PVDF membranes containing materials with hydrophilic groups exhibit a higher water transfer rate and lower filtration resistance than pure PVDF membranes, indicating their potential as high-performance separation membranes (Liu *et al.* 2011; Park *et al.* 2018; Matyjaszewski 2020). Amphiphilic polymers are branched macromolecules with branching periodicity and structural symmetry (Zhang & Dai 2019). These monodisperse branched macromolecules have numerous terminal groups on their surfaces, which increase their surface functionality and reactivity. During the phase inversion of membranes prepared using the homopolymer obtained by grafting an amphiphilic copolymer with PVDF, the hydrophilic chains of the amphiphilic copolymer migrate to the membrane surface, thus minimizing the interface free energy of the membrane. This is because the hydrophobic chains of amphiphilic copolymers are compatible with the main body of PVDF, whereas the hydrophilic chains achieve hydrophilicity through self-assembly and surface migration. Zhao *et al.* (2019) prepared polyacryloylmorpholine-b-poly(methyl methacrylate)-b-polyacryloylmorpholine (PAMA) using the reversible addition–fragmentation chain transfer polymerization method to fabricate a PVDF/PAMA UF membrane. The water flux, bovine serum albumin (BSA) retention, and recovery rate of the PVDF/PAMA membrane were 236 $L·m^{-2}·h^{-1}$, 98.3%, and 98.1%, respectively. Minehara *et al.* (2014) blended PVDF, methyl methacrylate, and polyethylene glycol methyl methacrylate copolymer (PMMA-g-PEO) to prepare a PVDF/PMMA-g-PEO membrane using the non-solvent induced phase separation (NIPS) method. The flux and rejection rate of polystyrene latex with a PVDF/PMMA-g-PEO membrane diameter of 88 nm were 29–50 $L·m^{-2}·h^{-1}$ and >96%, respectively. Wu *et al.* (2019) blended PVDF-grafted poly(ethylene glycol) methyl ether methacrylate (PEGMA) to prepare a PVDF/PVDF-g-PEGMA membrane. The flux, alginate rejection rate, and water flux recovery rate of the membrane were 700 $L·m^{-2}·h^{-1}$, 87%–100%, respectively, indicating the good antifouling of the membrane. These findings indicate that amphiphilic polymers can be used to improve the hydrophilic properties of PVDF UF membranes.

Ultraviolet (UV)-induced Cu(II)-mediated reversible deactivation radical polymerization (RDRP) is an environmentally friendly synthesis technology which combines UV stimulation and Cu(II)-mediated RDRP (Chuang *et al.* 2014; Hu *et al.* 2018a, 2018b). UV-induced Cu(II)-mediated RDRP can be used to modify pure PVDF membranes as it enables a more compact combination of hydrophobic and hydrophilic chains to ensure that the hydrophilic chain is fixed on the surface of the PVDF membranes (Cui *et al.* 2019; Zou *et al.* 2020; Li *et al.* 2021). Therefore, the addition of amphiphilic copolymers to PVDF membranes can improve the hydrophilicity and antifouling properties. In addition, the self-assembly capability of amphiphilic copolymers endows pure PVDF membranes with a higher permeation flux and narrow membrane pore size distribution (Guillen *et al.* 2011). Lei *et al.* (2018) used PVDF and tetrahydrofurfuryl methacrylate as raw materials to develop a new amphiphilic polymer membrane (THFMA, PVDF-g-THFMA) through photoinduced Cu(II)-mediated RDRP. The water flux and BSA rejection rate of the PVDF-g-THFMA copolymer UF membrane were 293.9 $L·m^{-2}·h^{-1}$ and 91.3%, respectively. Tong *et al.* (2020) prepare a novel amphiphilic copolymer of PVDF grafted with 1-Methyl-2-pyrrolidone (NMA, PVDF-g-NMA) using the same method. The water flux and BSA rejection rate of the PVDF-g-NMA copolymer UF membrane were 272.1 $L·m^{-2}·h^{-1}$ and 92.6%, respectively.

Papermaking wastewater has a complex composition and contains a large amount of lignin, hemicellulose, sugars, and other dissolved substances with high suspended solids, turbidity, and chemical oxygen demand (COD), which affect the survival of aquatic organisms. Generally, papermaking wastewater is treated using a combination of physical (e.g., PVDF UF using the membrane), chemical, and biological methods. The Ministry of Environmental Protection has recently stipulated that COD emissions are 80–100 mg/L, and the special emission limit is less than 50 mg/L. The papermaking enterprises are required to increase their efforts in the treatment of papermaking wastewater. Although previous studies have successfully improved the hydrophilicity of PVDF UF membranes, these membranes have not been applied in actual wastewater systems. N-Isobutoxy methacrylamide (IBMA) is a new energy-saving and environmentally friendly crosslinking monomer that can be used to improve the adhesion of polar substrates under UV radiation. In this work, we propose to prepare PVDF-g-IBMA

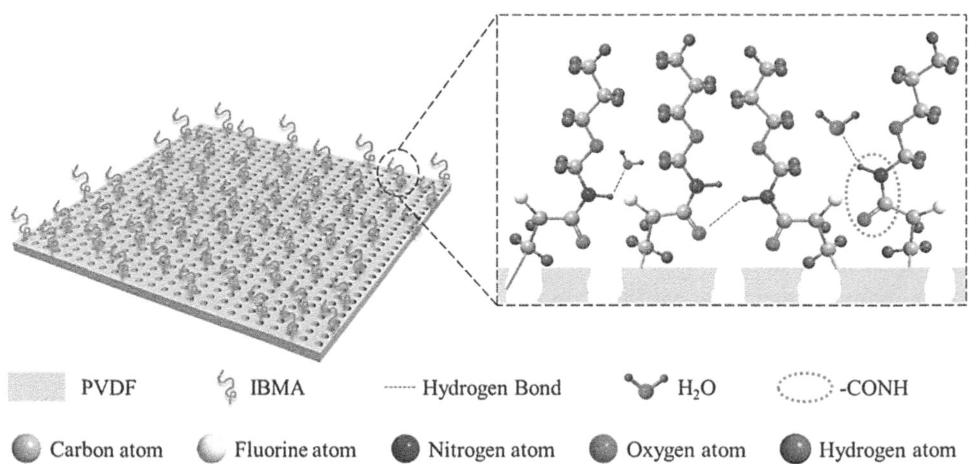

Figure 1 | The structure illustration of the PVDF-g-IBMA copolymer UF membrane.

copolymer by UV-induced Cu(II)-mediated RDRP using IBMA as a monomer as showed in Figure 1. We suppose the novel developed PVDF-g-IBMA membrane exhibits better hydrophilicity and antifouling properties than PVDF UF membranes, indicating its prospect for the treatment of papermaking wastewater.

2. EXPERIMENTAL SECTION

2.1. Chemicals

PVDF (Solef105, Mw 5,700,000–600,000) was obtained from Solvay Specialty Polymers (Shanghai, China). IBMA THFMA, dextran (Mw 10,000, 40,000, 70,000, and 500,000 Da), Me6-Tre, and BSA (Mw 67,000 Da) were bought from Sigma-Aldrich (Shanghai, China). $CuCl_2$, KCl, NaCl, $Na_2HPO_4 \cdot 12H_2O$, and KH_2PO_4, and NMP was procured from Aladdin Industrial Corporation (Shanghai, China).

2.2. Preparation of the PVDF-g-IBMA ultrafiltration membrane

PVDF (1.5 g), NMP (10 mL), and $CuCl_2$ (0.02 g) were added to a quartz reactor equipped with a rotor. The solution was purged with argon under stirring for 1 h to remove dissolved oxygen, after which Me6-Tren (0.1 mL) and IBMA solution (4.5 mL) were added to the solution (Figure 2). The obtained solution was polymerized by placing the solution in a UV reactor supplying 365 nm irradiation and protected under an argon atmosphere for 6 h. Subsequently, the obtained precipitate was washed with DI water three times to remove unreacted monomers. The obtained PVDF-g-IBMA was kept in an oven at 60 °C for further use.

PVDF-g-IBMA copolymer UF membranes were prepared using the NIPS method. The dried PVDF-g-IBMA copolymer was dissolved in a certain proportion of NMP solution in the three-necked flask (Table 1). The resulting casting solution was stirred in a 60 °C water bath with a mechanical stirrer with a speed of 0.2 kr/min for 4 h. The method of preparing the membrane

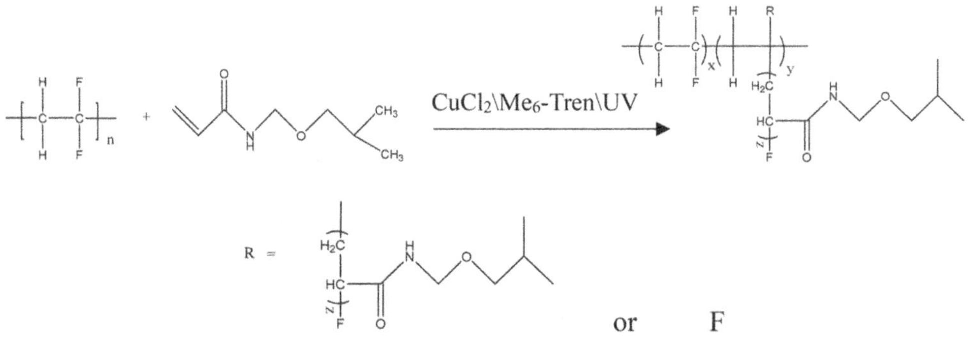

Figure 2 | Schematic diagram of grafting procedure between PVDF and IBMA.

Table 1 | Membrane prepared with different ratios

Membrane	PVDF (wt.%)	PVDF-g-IBMA (wt.%)	NMP (wt.%)
P15	15	/	85
P20	20	/	80
P25	25	/	75
I15	/	15	85
I20	/	20	80
I25	/	25	75

is the same as previously reported. The viscosity of the UF membrane casting solution was tested by selecting a suitable rotor and rotation speed.

2.3. Characterization of PVDF-g-IBMA copolymer UF membrane

The grafting reaction of the PVDF-g-IBMA copolymer was investigated using ^{1}H nuclear magnetic resonance (NMR, AVANCE AV-300, Bruker, Switzerland) spectroscopy. ^{1}H NMR spectroscopy sample was prepared by dissolving the PVDF-g-IBMA copolymer in deuterated dimethyl sulfoxide. The mass and molar fractions of IBMA in the PVDF-g-IBMA copolymer were calculated using Equations (1) and (2), respectively.

$$\varphi_w(IBMA) = \frac{\varphi_{m(IBMA)} \cdot M_{IBMA}}{\varphi_{m(IBMA)} \cdot M_{IBMA} + (1 - \varphi_{m(IBMA)}) \cdot M_{PVDF(unit)}} \tag{1}$$

$$\varphi_m(IBMA) = \frac{\frac{1}{15}(I_a + I_b + I_c + I_d + I_e + I_f)}{\frac{1}{15}(I_a + I_b + I_c + I_d + I_e + I_f) + \frac{1}{2}(I_{a(hh)} + I_{a(ht)})}, \tag{2}$$

where $\varphi_w(IBMA)$ and $\varphi_m(IBMA)$ are the mass and mole fractions of IBMA in PVDF-g-IBMA, respectively; I_X is the area of the corresponding peak in the ^{1}H NMR spectrum; and M_{IBMA} and $M_{PVDF(unit)}$ are the molecular weights of IBMA and PVDF, respectively.

The Fourier-transform infrared Spectrometer (FTIR) of the UF membranes were analyzed using a FTIR spectrometer (Cary660, Agilent, Australia) by attenuated total reflectance in the wavelength range of 4,000–500 cm^{-1}. The melting points of the PVDF and PVDF-g-IBMA copolymers were measured using differential scanning calorimetry (DSC, STA449F3, NETZSCH, Germany). SEM images of all the PVDF-g-IBMA UF membranes were attained using a scanning electron microscopy (SEM, Hitachi S4800, Japan) instrument. To examine the chemical compositions of copolymer membranes, the distributions of oxygen, carbon, and fluorine in the membrane were recorded using SEM equipped with an energy-dispersive X-ray spectrometer. The dynamic contact angles (DWCA) of the UF membranes were analyzed using a contact angle measuring instrument (DSA 100, Kruss, Germany). The WCAs were tested by placing a 3.0 μL DI water droplet on the membrane surfaces.

2.4. Separation performance

The pure water fluxes of the PVDF homopolymer and PVDF-g-IBMA copolymer UF membranes with an effective area (4.1 cm^2) were measured using an UF apparatus (Amicon 8400). For the analysis, first, the membranes were pre-compacted at 0.15 MPa for 20 min. Subsequently, the pure water fluxes of the UF membranes were recorded at 0.1 MPa for 5 min. The pure water fluxes of the UF membranes were calculated using Equation (3).

$$J = \frac{V}{A \, \Delta t} \tag{3}$$

where V is the volume of pure water permeated through the membrane (L), A is the effective surface area of the membrane (m^2), and Δt is the permeation time (h).

To determine the retention performance of the modified membrane, dextran solution was used as the filter medium by the method reported previously (Tong *et al.* 2020). Molecular weight cutoff (MWCO) is the membrane's retention rate of 90% for dextran solution. Gel chromatography is used to analyze the molecular weight distribution of the solution on the feed side and the permeate side.

The antifouling performance of the PVDF homopolymer and PVDF-g-IBMA copolymer membranes was measured using the apparatus at 0.1 MPa for 60 min (Tong *et al.* 2020). The protein fouling potential of the PVDF-g-IBMA UF membranes was evaluated using a BSA solution. To prepare the BSA solution, first, PBS solution (pH 7.4) was prepared by adding NaCl (8.0 g), $Na_2HPO_4 \cdot 12H_2O$ (3.63 g), KCl (0.2 g), and KH_2PO_4 (0.24 g) to DI water (1.0 L). Subsequently, BSA (0.5 g) was mixed with the above solution to a concentration of 0.5 g/L.

The BSA rejection rate (R) of the prepared UF membranes was calculated using Equation (4).

$$R(\%) = \left(1 - \frac{C_p}{C_f}\right) \times 100\%, \tag{4}$$

where R is the BSA rejection rate, and C_p and C_f are the protein concentrations in the permeate and feed (g/L), respectively. The C_p and C_f values were resolved using a Ultraviolet-visible Spectrophotometer (UV-6000, Shanghai Yuanyi Apparatus Co., Ltd, China) in the wavelength of 280 nm. The calibration curve of BSA concentration was consistent with those in previous reports (Tong *et al.* 2020).

The tested membranes were rinsed repeatedly with deionized water. Subsequently, the pure water recovery ratios of the prepared UF membranes were calculated using Equation (5).

$$FRR(\%) = \frac{J_r}{J_w} \times 100\% \tag{5}$$

where FRR (%) is the pure water recovery ratio, and J_r and J_w are the original pure and recovered water fluxes $(L \cdot m^{-2} \cdot h^{-1})$ after the washing procedure, respectively.

2.5. Performance of membrane separation industrial papermaking wastewater

The industrial wastewater was obtained from reverse osmosis concentrated water from a paper mill in Suzhou. The wastewater is tested for pH, COD, turbidity, conductivity, and various anions and cations (Table 2).

The laboratory self-made cross-flow device was used for reverse osmosis (RO) concentrated water filtration experiments. PVDF UF membrane and PVDF-g-IBMA UF membrane were selected to test their performance. The pressure was set to 0.1 MPa. The effective area of the device is 2.83 cm^2. The experimental measurement time was set to 120 min.

3. RESULTS AND DISCUSSION

3.1. Characterization of PVDF-g-IBMA copolymer UF membrane

3.1.1. ^{1}H NMR

The structures of the PVDF homopolymer and PVDF-g-IBMA copolymer were verified using ^{1}H NMR spectroscopy (Figure 3 and Table 3). The peaks observed at 2.25 and 2.89 ppm in the ^{1}H NMR spectra of the copolymers corresponded to the $-CF_2CH_2CH_2CF_2-$ (head-to-head, hh) and $-CH_2CF_2CH_2CF_2-$ (head-to-tail, ht) of PVDF, respectively. In addition, the

Table 2 | RO concentrated water quality monitoring

Parameter	Unit	Value	Parameter	Unit	Value
COD	mg/L	720	Cu^{2+}	mg/L	79.7
Turbidity	NTU	1.68	K^+	mg/L	0
Conductivity	ms/cm	4.74	Ca^{2+}	mg/L	626.8
pH	/	7.20	Na^+	mg/L	1,367.0
SO_4^{2-}	mg/L	464.7	Fe^{2+}	mg/L	0.1
Cl^-	mg/L	973.2	Al^{3+}	mg/L	0
NO_3^-	mg/L	8.1	Mg^{2+}	mg/L	168.4

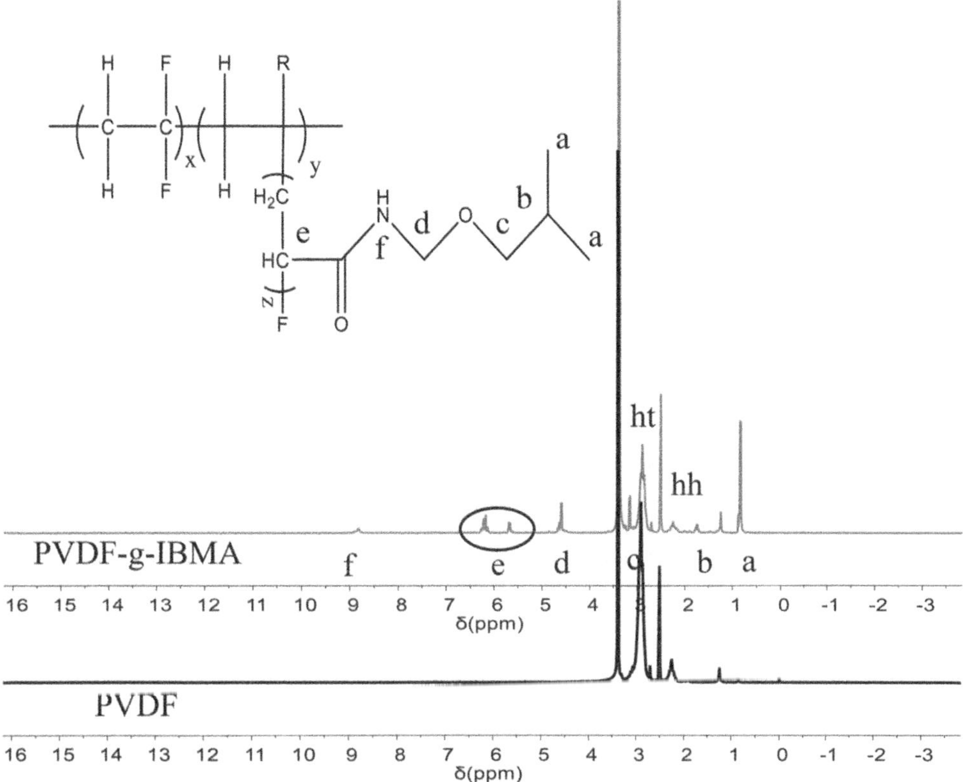

Figure 3 | The ^{1}H NMR spectra of the PVDF homopolymer and PVDF-g-IBMA copolymer.

Table 3 | The peak area of PVDF-g-IBMA copolymer in ^{1}H NMR spectra

I_x	I_a	I_b	I_c	I_d	I_e	I_f	I_{hh}	I_{ht}
1.00	0.18	0.29	0.47	0.55	0.17	0.58	3.84

characteristic peaks of methyl, methane, and N–H were observed at $\delta = 0.91$ ppm (a), $\delta = 1.24$ ppm (b), and $\delta = 8.82$ ppm (f), respectively. The remaining peaks observed in the ^{1}H NMR spectra were the peaks of CH_2. Furthermore, extra resonances were inspected in the ^{1}H NMR spectrum of the PVDF-g-IBMA copolymer, which corresponded to the protons in IBMA, verifying the existence of IBMA in the PVDF-g-IBMA copolymers. In addition, the molar and mass fractions of IBMA were 7.43% and 16.47%, respectively.

3.1.2. FTIR spectroscopy

FTIR spectroscopy was carried out to verify the synthesis of PVDF-g-IMBA copolymer UF membrane. Several peaks were observed in the FTIR of the PVDF-g-IMBA UF membrane compared with that of the PVDF UF membrane (Figure 4). The peaks observed at 1,541 and 1,667 cm^{-1} could be ascribed to the bending vibrations of C–N–H and C$=$O, respectively, whereas those at 2,830–2,940 cm^{-1} could be assigned to the stretching vibrations of CH_2 and CH_3. These observations confirmed the successful synthesis of PVDF-g-IBMA copolymer.

3.1.3. DSC analysis

The melting point of the PVDF-g-IBMA copolymer was analyzed using DSC analysis. The melting point of the PVDF-g-IBMA copolymer was 174.2 °C, which was higher than that of the PVDF homopolymer (170.2 °C) (Figure 5). The increase in the melting point of PVDF corresponded to the change in the crystal form of the PVDF-g-IBMA copolymer. Thermogravimetric analysis further confirmed the successful grafting of IBMA onto PVDF.

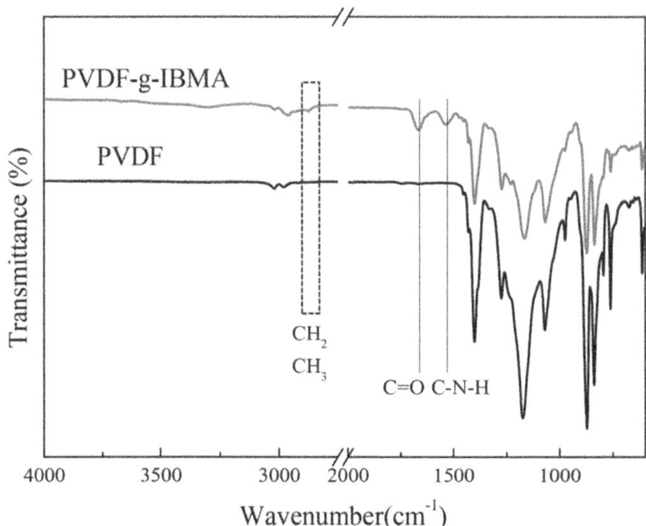

Figure 4 | FT-IR/ATR spectra of PVDF homopolymer and PVDF-g-IBMA copolymer UF membrane.

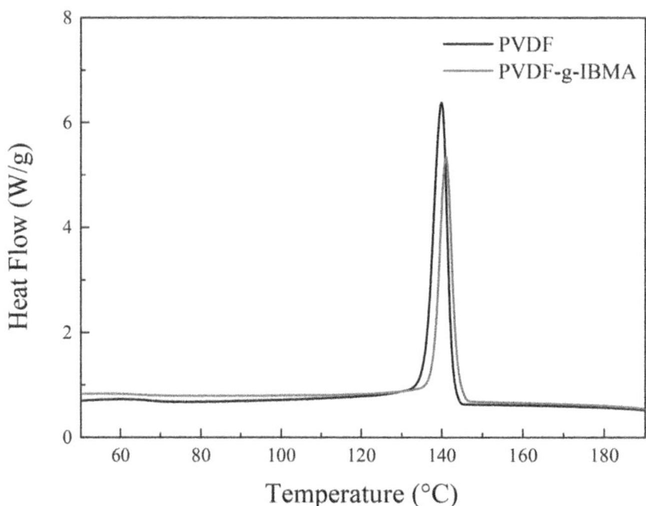

Figure 5 | DSC analysis of PVDF homopolymer and PVDF-g-IBMA copolymer UF membrane.

3.1.4. Casting solution viscosity

As shown in Figure 6, the viscosity of PVDF-g-IBMA UF casting membrane solution increased with an increase in the amount of the PVDF-g-IBMA copolymer in the casting solution, in the order of I15 < I20 < I25. In addition, the viscosity of PVDF-g-IBMA copolymer casting solution was higher than that of pure PVDF homopolymer casting solution with the same polymer concentration. An increase in the viscosity of the membrane casting solution hindered the exchange of the solvent and non-solvent during UF membrane formation and affected phase separation. However, the PVDF-g-IBMA copolymer contained a hydrophilic amide group (–NHCO–) that increased the phase rate of its membrane formation. These results indicated that the viscosity and hydrophilic chains of the casting solution affected the phase conversion rate during membrane formation, thus affecting the structure and performance of the prepared membrane owing to the competition between the two factors.

3.1.5. Membrane hydrophilicity

To illustrate the surface wettability of the PVDF-g-IBMA copolymer UF membrane, the DWCAs of the PVDF-g-IBMA UF membranes were measured and compared with those of the pure PVDF UF membranes (Figure 7). Within 180 s, the CA of the PVDF-g-IBMA copolymer membrane surface decreased. This was because penetration is a dynamic process, and

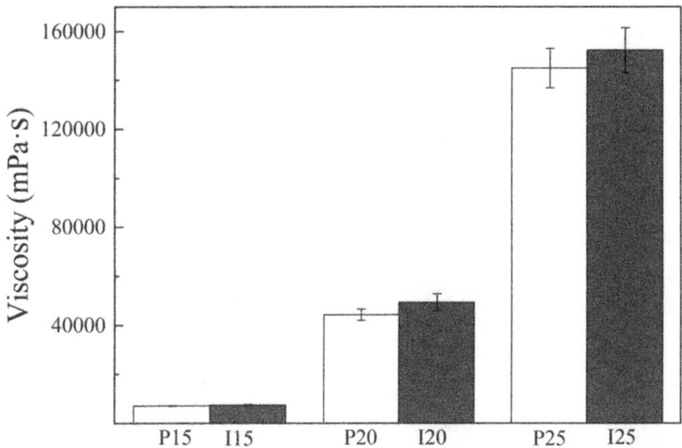

Figure 6 | Casting solution viscosity of the PVDF homopolymer and PVDF-g-IBMA copolymer casting solution.

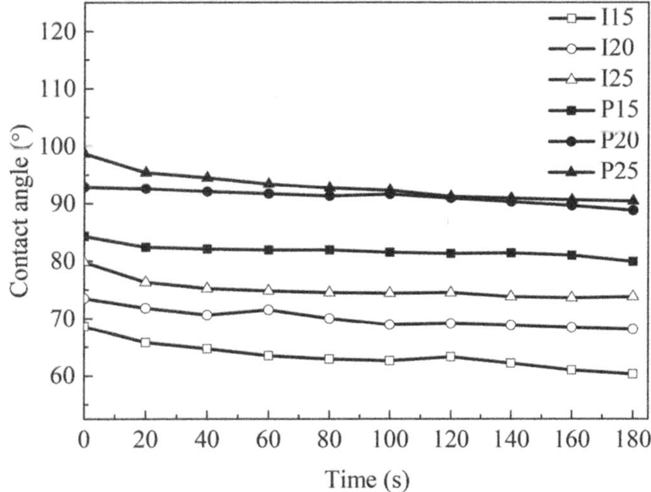

Figure 7 | DCA of the PVDF homopolymer and PVDF-g-IBMA copolymer UF membrane.

the CA of the membrane surface changed over time. In addition, the initial CAs of the pure PVDF UF membranes were larger than those of the PVDF-g-IBMA UF membrane. After 180 s, the stable CAs of I15, I20, and I25 were 60.3°, 68.1°, and 73.8°, respectively, which were less than those of the pure PVDF UF membranes. This indicated that the PVDF-g-IBMA UF membranes exhibited better hydrophilic performance than the PVDF membranes. Moreover, the surface of the PVDF-g-IBMA UF membrane was rich in IBMA hydrophilic chains. The hydrophilic amide groups in IBMA formed hydrogen bonds with H_2O molecules and formed a water layer, which isolated pollutants. The addition of hydrophilic groups changes the hydrophilic properties of the membrane. However, as the concentration of hydrophilic substances increases, the migration resistance became larger. As a result, there are more hydrophilic groups on the surface of the membrane at lower concentration, which leads to the decrease of the contact angle.

3.1.6. SEM analysis

As shown in Figure 8, the surface and cross-sectional morphologies images of the PVDF-g-IBMA copolymer UF membranes were investigated using SEM. The surface morphology characteristics of the PVDF-g-IBMA UF membranes were similar to those of the pure PVDF UF membranes (Tong *et al.* 2020). As the concentration of the membrane casting solution increased, the number of pores on the surface of the formed UF membrane decreased. Consequently, the surface of I15 had the most abundant pores. In addition, notable pore structures and sponge layers were observed in the cross-sectional morphology

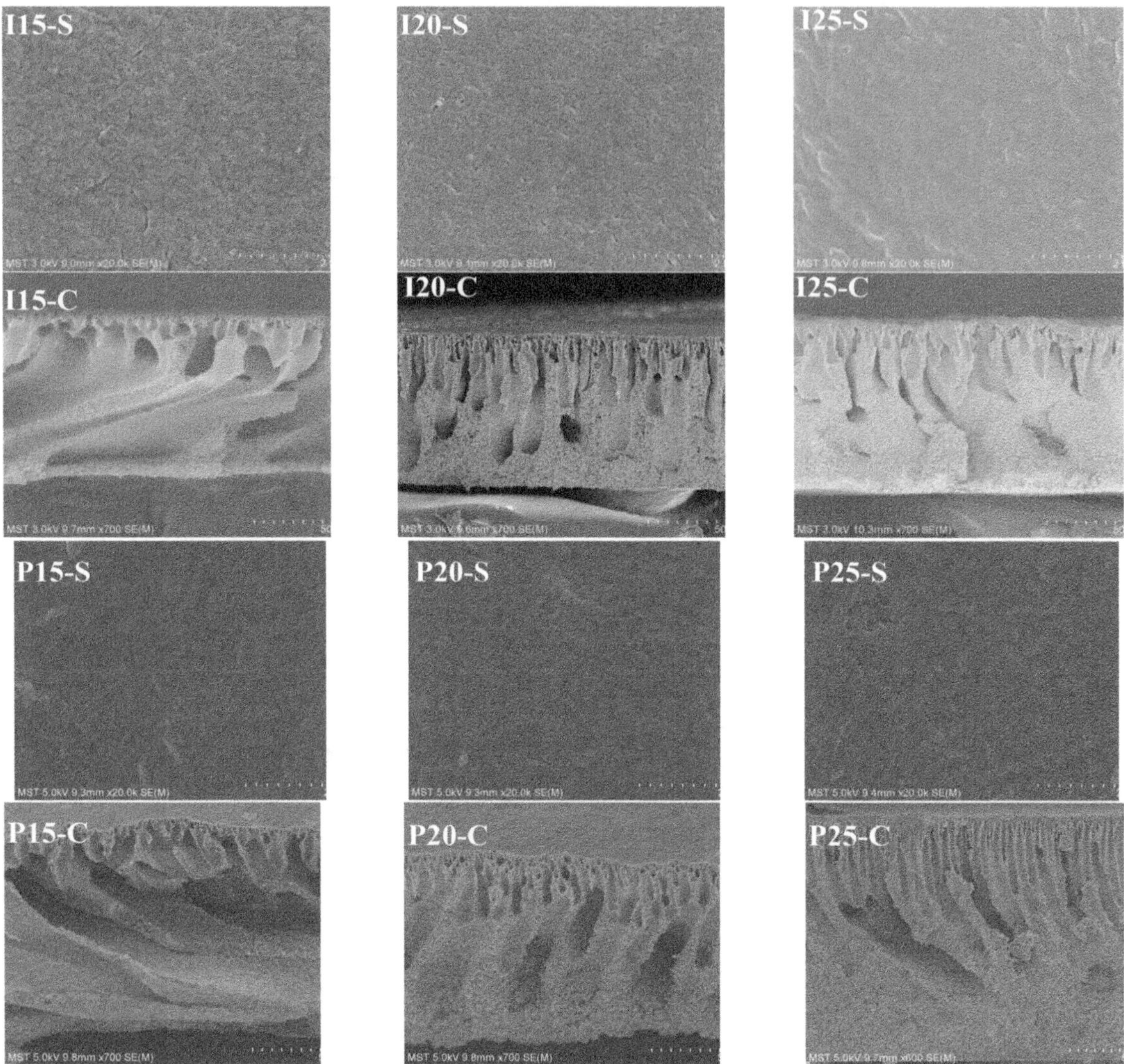

Figure 8 | SEM images of the surface and cross-section morphologies of the PVDF-g-IBMA copolymer UF membrane and the PVDF homopolymer.

of the PVDF-g-IBMA UF membranes. Because the casting solution of I15 had the lowest viscosity, it exhibited the highest phase inversion rate during phase inversion, which resulted in the formation of notable finger-like macroporous structures. As the concentration of casting solution increased, the number of finger-like pores gradually decreased, and the area of the sponge layer gradually increased. Moreover, the SEM image revealed the successful formation of membranes with finger-like macropores due to the rapid exchange rate of the solvent and non-solvent.

3.1.7. EDX analysis

The distribution of elements in the cross-section of the PVDF-g-IBMA copolymer UF membrane is shown in Figure 9. After scanning the cross-sectional energy spectrum, in addition to the carbon and fluorine atoms of PVDF, the oxygen and nitrogen atoms of IBMA were observed, confirming the successful synthesis of the PVDF-g-IBMA copolymer. Notable pore structures and sponge layers were observed at the upper and lower parts of the cross-section, respectively. However, the distribution ratio of nitrogen atoms in the upper and lower parts of the cross-section was 1:1, indicating that the hydrophilic chains of IBMA migrated to surface of pores (Oikonomoua *et al.* 2017).

3.1.8. AFM analysis

The AFM images of the membrane surface on a scanning area of 5 μm by 5 μm are shown in Figure 10. The average roughness of the PVDF-g-IBMA membrane is much lower than that of the PVDF membrane (P15 = 147 nm, P20 = 267 nm,

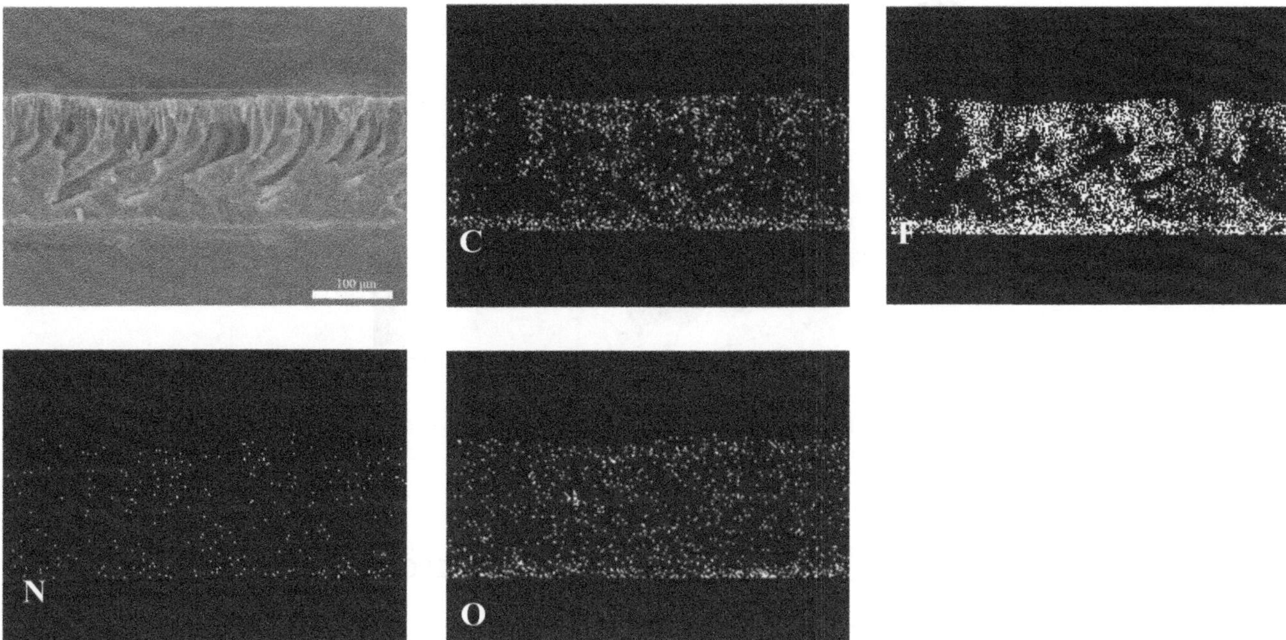

Figure 9 | EDX mapping of the PVDF-g-IBMA copolymer UF membrane.

P25 = 129 nm) (Tong *et al.* 2020). The surface roughness of I20 is 39.8 nm, while it is 147 nm for P15. Therefore, the PVDF-g-IBMA membrane with a smoother surface may have greater antifouling ability.

3.1.9. Mechanical properties

In order to investigate the mechanical properties of PVDF and PVDF-g-IBMA membrane, the tensile strength and elongation at break values were tested. As listed in Table 4, compared with the PVDF membrane, the tensile strength and elongation at break of PVDF-g-IBMA membrane were reduced. It indicated that the mechanical properties of the PVDF-g-IBMA membrane reduced a little.

3.2. Membrane performance

3.2.1. Pure water flux and MWCO

The pure water fluxes of the PVDF and PVDF-g-IBMA copolymer UF membranes are shown in Figure 11. As the concentration of casting solution increased, the pure water fluxes of the UF membranes decreased. The pure water fluxes of the PVDF-g-IBMA UF membranes were significantly higher than those of the pure PVDF UF membranes. The pure water flux of I15 was 432.8 L·m^{-2}·h^{-1}, which was higher than those of the other PVDF-g-IBMA UF membranes and 7.5 times higher than that of P15. The pure water fluxes of I20 and I25 were 7.4 and 23.8 times higher than those of P20 and P25, respectively. These results indicated that the PVDF-g-IBMA membranes had denser pores than the pure PVDF UF membranes, and the migration effect of hydrophilic chains to the surface of the PVDF-g-IBMA UF membranes was more notable.

The MWCO values of I15, I20, and I25 were 43.7, 23.2, and 16.6 kDa, respectively. The MWCO values of the PVDF-g-IBMA UF membranes were higher than those of the pure PVDF UF membranes of the same concentration. These results were consistent with the microporous structure and higher pure water fluxes of the PVDF-g-IBMA UF membranes.

3.2.2. Anti-fouling performance of the PVDF-g-IBMA copolymer UF membrane

The BSA rejection and flux recovery rates of the PVDF-g-IBMA copolymer UF membranes are listed in Table 5. The BSA rejection rates of the PVDF-g-IBMA UF membranes were higher than those of the PVDF UF membranes. The BSA rejection rate of I15 (88.4%) was higher than that of P15 (85.7%). As the concentration of the casting solution for the PVDF-g-IBMA membranes increased, the BSA rejection rate of the copolymer UF membranes increased. Consequently, I25 had the highest BSA rejection rate (92.7%). Furthermore, the pure water flux recovery rates of the PVDF-g-IBMA UF membranes were higher

I15

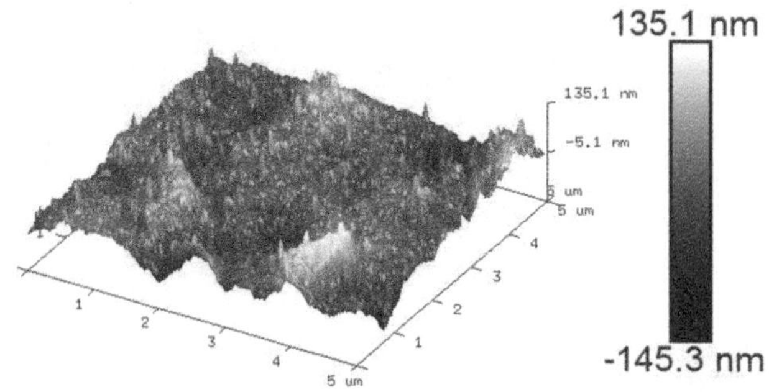

I20

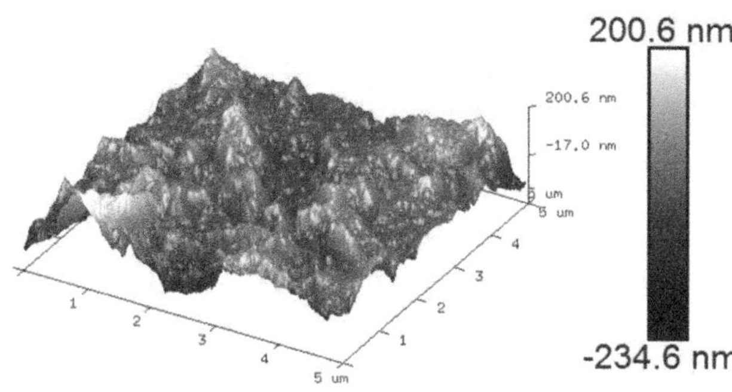

I25

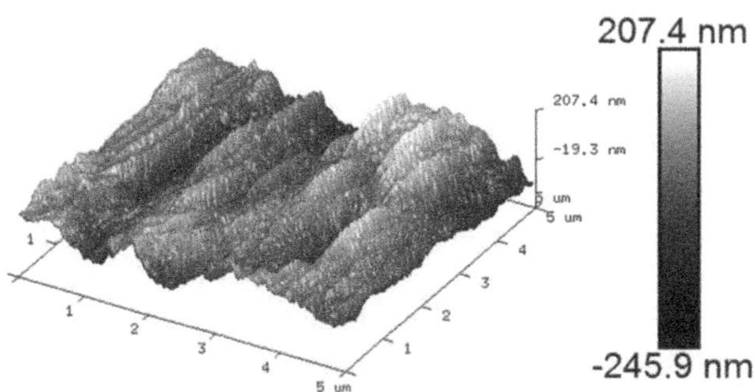

Figure 10 | 3D AFM images of the PVDF and PVDF-g-IBMA membrane surfaces.

than those of the pure PVDF UF membranes. The pure water flux recovery rate of I25 was 89.5%, which was significantly higher than that of the pure PVDF UF membrane (P20, 78.0%).

As the concentration of the copolymer casting solution increased, the BSA initial and stable fluxes of the PVDF-g-IBMA UF membranes decreased (Figure 12). The stable fluxes of the PVDF-g-IBMA UF membranes were higher than those of the pure PVDF UF membranes. The stable flux of I15 (164.6 L·m^{-2}·h^{-1}) was significantly higher than that of P15 (27.8 L·m^{-2}·h^{-1}). The stable fluxes of I20 and 125 were significantly higher than those of P20 and P25. These results corresponded to the improved antifouling of the PVDF-g-IBMA UF membranes compared with those of the PVDF membranes (Tong *et al.* 2020).

Table 4 | Mechanical properties of PVDF and PVDF-g-IBMA membrane

	Elongation at break values (%)	Tensile strength (MPa)
P15	18.31	1.94
P20	20.04	2.42
P25	32.31	3.74
I15	15.30	1.44
I20	18.75	1.90
I25	26.11	3.13

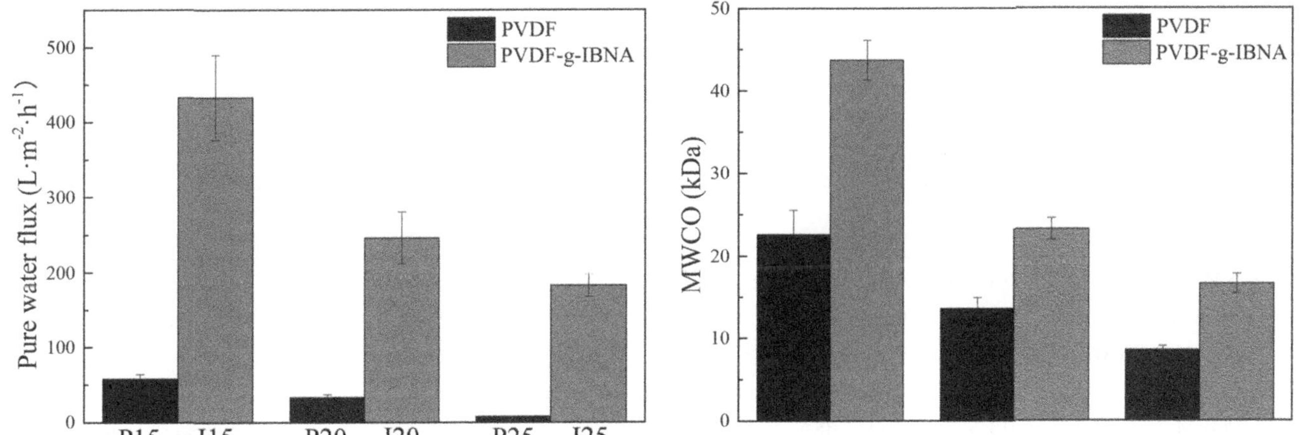

Figure 11 | Pure water flux and MWCO of the PVDF-g-IBMA copolymer UF membrane.

Table 5 | BSA rejection and FRR of the PVDF-g-IBMA copolymer UF membrane

Membrane code	BSA rejection(%)	FRR(%)
I15	88.4 ± 2.7	90.8 ± 3.1
I20	90.5 ± 3.8	92.3 ± 2.6
I25	92.7 ± 1.6	89.5 ± 2.5
P15 (Tong *et al.* 2020)	85.7 ± 3.4	79.0 ± 2.1
P20 (Tong *et al.* 2020)	86.4 ± 2.2	82.9 ± 2.4
P25 (Tong *et al.* 2020)	88.9 ± 2.6	78.0 ± 3.6

Compared with published modified PVDF membranes, including PVDF-g-NMA, PVDF-g-PEGMA, PVDF-g-POEM, PVDF-g-PMABS, PVDF-g-(PAMCO-PAA), The pure water flux of I15 ($432.8 \text{ L·m}^{-2}\text{·h}^{-1}$) was significantly higher than those of the previously modified copolymer membranes (Table 6).

3.3. Treatment of RO-concentrated water using the PVDF-g-IBMA copolymer ultrafiltration membrane

The performances of the pure PVDF and PVDF-g-IBMA UF membranes for the treatment of RO-concentrated water are shown in Table 7 and Figure 13. The COD and turbidity removal rates of the PVDF-g-IBMA UF membranes were significantly improved compared with those of the pure PVDF UF membranes. Under the same casting solution concentration, the actual wastewater treatment effect of the PVDF-g-IBMA copolymer UF membranes was stronger than those of the PVDF UF membranes. The COD and turbidity removal rates of I25 were 61.5% and 92.8%, respectively, which were higher than those of the

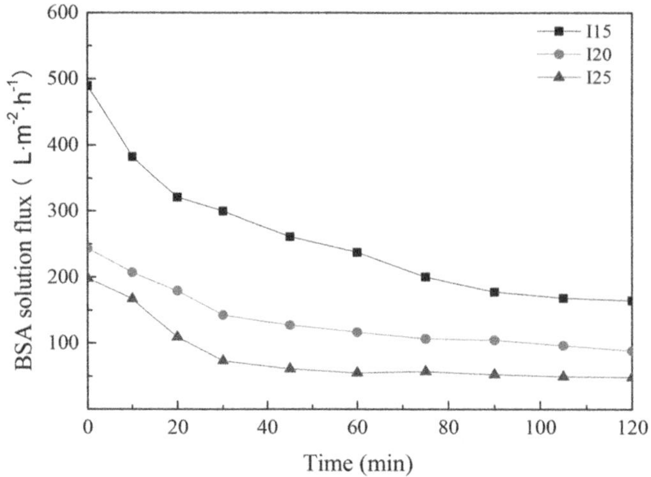

Figure 12 | Time-dependent BSA solution flux of the PVDF-g-IBMA copolymer UF membrane.

Table 6 | Comparison with the results of literatures

Membrane	Operating conditions	Pure water (L·m⁻²·h⁻¹)	Rejection (%)	References
PVDF-g-NMA	0.1 MPa, 1.0 g·L⁻¹BSA	272	93.7	Tong *et al.* (2020)
PVDF-g-PMABS	0.1 MPa, 1.0 g·L⁻¹BSA	136	98.6	Chen *et al.* (2018)
PVDF-g-(PAMCO-PAA)	0.1 MPa, 1.0 g·L⁻¹BSA	130	96.3	Xu *et al.* (2017)
PVDF-g-PEGMA	0.1 MPa, 1.0 wt.% BSA	116	66.3	Hashim *et al.* (2009)
PVDF-g-POEM	0.1 MPa, 1.0 g·L⁻¹BSA	130	83.5	Moghareh Abed *et al.* (2013)
This work	0.1 MPa, 1.0 g·L⁻¹BSA	432	88.4	

Table 7 | Water quality of liquid penetration after filtering RO concentrated water with PVDF-g-IBMA copolymer UF membrane

Membrane	Permeate side COD (mg/L)	COD removal rate (%)	Permeate turbidity (NTU)	Turbidity removal rate (%)
P15	600	16.7	0.176	89.5
P20	510	29.2	0.161	90.4
P25	470	34.7	0.159	90.5
I15	419	41.8	0.165	90.2
I20	365	49.3	0.140	91.7
I25	277	61.5	0.121	92.8

other samples. The RO concentrated water flux of the membranes decreased over time. The hydrophilicity of the PVDF-g-IBMA copolymer UF membranes was better than that of the PVDF UF membranes. Consequently, the initial and stable fluxes of the PVDF-g-IBMA copolymer UF membranes were significantly higher than those of the PVDF UF membranes.

4. CONCLUSION

A novel PVDF-g-IBMA amphiphilic copolymer was successfully synthesized via ultraviolet-induced Cu(II)-mediated RDRP and used to fabricate a UF membrane using the NIPS method. ¹H NMR, FTIR spectroscopy, and DSC analysis confirmed the successful synthesis of the PVDF-g-IBMA copolymer. The hydrophilicity and antifouling performance of the prepared PVDF-g-IBMA copolymer UF membranes were improved compared with those of the pure PVDF UF membranes. The contact angle of the UF membrane reduced to 60.3° after 180 s, and the pure water flux of the UF membrane was

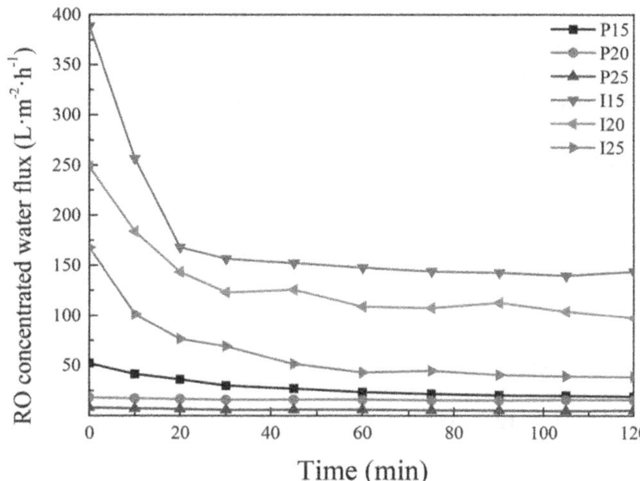

Figure 13 | Time-dependent RO concentrated water fluxes of PVDF homopolymer UF membrane and PVDF-g-IBMA copolymer UF membrane.

432.8 L·m^{-2}·h^{-1}, which was 7.5 times higher than that of the pure PVDF UF membrane. The MWCO, BSA rejection rate, and pure water flux of the UF membrane were 43.7 kDa, 88.4%, and 90.8%, respectively. The stable flux of the PVDF-g-IBMA UF membrane for filtrating BSA solution was 164.6 L·m^{-2}·h^{-1}, which was higher than that of the pure PVDF UF membrane (27.8 L·m^{-2}·h^{-1}).

For the treatment of papermaking wastewater, the COD was reduced from 720 to 277 mg/L with a removal rate of 61.5%. The turbidity decreased from 1.68 to 0.12 NTU with a removal rate of 92.8%. And the initial and stable fluxes of the PVDF-g-IBMA copolymer UF membranes were significantly higher than those of the pure PVDF UF membranes, indicating the potential of prepared membrane for treating papermaking wastewater.

ACKNOWLEDGEMENTS

This work was supported by Tianjin Synthetic Biotechnology Innovation Capacity Improvement Project (TSBICIP-KJGG-003), Qinglan Plan of Jiangsu Education Department and the National Key R&D Program of China (2017YFD0400402).

DECLARATION OF COMPETING INTERESTS

The authors declare that they have no known competing financial interests or personal relationships that could have appeared to influence the work reported in this paper.

CREDIT AUTHORSHIP CONTRIBUTION STATEMENT

Yujia Tong: Conceptualization, Methodology, Software, Formal analysis, Investigation, Writing – Original Draft, Writing – Review and Editing. **Wenlong Ding:** Validation, Investigation. **Lijian Shi:** Validation, Investigation. **Weixing Li:** Writing – Review and Editing, Supervision, Funding acquisition.

DATA AVAILABILITY STATEMENT

All relevant data are included in the paper or its Supplementary Information.

REFERENCES

Chen, F., Shi, X., Chen, X. & Chen, W. 2018 Preparation and characterization of amphiphilic copolymer PVDF-g-PMABS and its application in improving hydrophilicity and protein fouling resistance of PVDF membrane. *Applied Surface Science* **427**, 787–797.
Chen, M., Ding, W., Zhou, M., Zhang, H., Ge, C., Cui, Z. & Xing, W. 2021 Fouling mechanism of PVDF ultrafiltration membrane for secondary effluent treatment from paper mills. *Chemical Engineering Research and Design* **167**, 37–45.

Chuang, Y.-M., Ethirajan, A. & Junkers, T. 2014 Photoinduced sequence-controlled copper-mediated polymerization: synthesis of decablock copolymers. *ACS Macro Letters* **3** (8), 732–737.

Cui, Z., Tang, X., Li, W., Liu, H., Zhang, J., Wang, H. & Li, J. 2019 EVOH in situ fibrillation and its effect of strengthening, toughening and hydrophilic modification on PVDF hollow fiber microfiltration membrane via TIPS process. *Journal of Materials Science* **54** (7), 5971–5987.

Gao, W., Liang, H., Ma, J., Han, M., Chen, Z.-l., Han, Z.-s. & Li, G.-b. 2011 Membrane fouling control in ultrafiltration technology for drinking water production: a review. *Desalination* **272** (1–3), 1–8.

Gao, Y., Qin, J., Wang, Z. & Østerhus, S. W. 2019 Backpulsing technology applied in MF and UF processes for membrane fouling mitigation: a review. *Journal of Membrane Science* **587**, 117136.

Gönder, Z. B., Arayici, S. & Barlas, H. 2012 Treatment of pulp and paper mill wastewater using utrafiltration process: optimization of the fouling and rejections. *Industrial & Engineering Chemistry Research* **51** (17), 6184–6195.

Guillen, G. R., Pan, Y., Li, M. & Hoek, E. M. V. 2011 Preparation and characterization of membranes formed by nonsolvent induced phase separation: a review. *Industrial & Engineering Chemistry Research* **50** (7), 3798–3817.

Haq, I., Mazumder, P. & Kalamdhad, A. S. 2020 Recent advances in removal of lignin from paper industry wastewater and its industrial applications – a review. *Bioresour Technol* **312**, 123636.

Hashim, N. A., Liu, F. & Li, K. 2009 A simplified method for preparation of hydrophilic PVDF membranes from an amphiphilic graft copolymer. *Journal of Membrane Science* **345**, 134–141.

Hu, X., Cui, G., Zhang, Y., Zhu, N. & Guo, K. 2018a Copper(II) photoinduced graft modification of P(VDF- co -CTFE). *European Polymer Journal* **100**, 228–232.

Hu, X., Cui, G., Zhu, N., Zhai, J. & Guo, K. 2018b Photoinduced Cu(II)-Mediated RDRP to P(VDF-co-CTFE)-g-PAN. *Polymers (Basel)* **10** (1), 68.

Lei, H., Liu, L., Huang, L., Li, W. & Xing, W. 2018 Novel anti-fouling PVDF-g-THweFMA copolymer membrane fabricated via photoinduced Cu(II)-mediated reversible deactivation radical polymerization. *Polymer* **157**, 1–8.

Li, R., Li, J., Rao, L., Lin, H., Shen, L., Xu, Y., Chen, J. & Liao, B.-Q. 2021 Inkjet printing of dopamine followed by UV light irradiation to modify mussel-inspired PVDF membrane for efficient oil-water separation. *Journal of Membrane Science* **619**, 118790.

Liu, F., Hashim, N. A., Liu, Y., Abed, M. R. M. & Li, K. 2011 Progress in the production and modification of PVDF membranes. *Journal of Membrane Science* **375** (1–2), 1–27.

Matyjaszewski, K. 2020 Amphiphilic polymer co-networks: synthesis, properties, modelling and applications. Edited by Costas S. Patrickios. *Angewandte Chemie International Edition* **60** (3), 1064–1064.

Minehara, H., Dan, K., Ito, Y., Takabatake, H. & Henmi, M. 2014 Quantitative evaluation of fouling resistance of PVDF/PMMA-g-PEO polymer blend membranes for membrane bioreactor. *Journal of Membrane Science* **466**, 211–219.

Moghareh Abed, M. R., Kumbharkar, S. C., Groth, A. M. & Li, K. 2013 Economical production of PVDF-g-POEM for use as a blend in preparation of PVDF based hydrophilic hollow fibre membranes. *Separation and Purification Technology* **106**, 47–55.

Oikonomoua, E., Karpati, S., Gassara, S., Deratani, A., Beaume, F., Lorain, O., Tencé-Girault, S. & Norvez, S. 2017 Localization of antifouling surface additives in the pore structure of hollow fiber PVDF membranes. *Journal of Membrane Science* **538**, 77–85.

Park, S.-H., Ahn, Y., Jang, M., Kim, H.-J., Cho, K. Y., Hwang, S. S., Lee, J.-H. & Baek, K.-Y. 2018 Effects of methacrylate based amphiphilic block copolymer additives on ultra filtration PVDF membrane formation. *Separation and Purification Technology* **202**, 34–44.

Shen, Z., Cai, N., Xue, Y., Yu, B., Wang, J., Song, H., Deng, H. & Yu, F. 2020 Porous SBA-15/cellulose membrane with prolonged anti-microbial drug release characteristics for potential wound dressing application. *Cellulose* **27** (5), 2737–2756.

Sierke, J. & Ellis, A. V. 2019 Cross-linking of dehydrofluorinated PVDF membranes with thiol modified polyhedral oligomeric silsesquioxane (POSS) and pure water flux analysis. *Journal of Membrane Science* **581**, 362–372.

Sri Abirami Saraswathi, M. S., Rana, D., Divya, K., Alwarappan, S. & Nagendran, A. 2018 Fabrication of anti-fouling PVDF nanocomposite membranes using manganese dioxide nanospheres with tailored morphology, hydrophilicity and permeation. *New Journal of Chemistry* **42** (19), 15803–15810.

Sun, C.-C., Zhou, M.-Y., Yuan, J.-J., Yan, Y., Song, Y.-Z., Fang, L.-F. & Zhu, B.-K. 2020 Membranes with negatively-charged nanochannels fabricated from aqueous sulfonated polysulfone nanoparticles for enhancing the rejection of divalent anions. *Journal of Membrane Science* **602**, 117692.

Toczyłowska-Mamińska, R. 2017 Limits and perspectives of pulp and paper industry wastewater treatment – A review. *Renewable and Sustainable Energy Reviews* **78**, 764–772.

Tong, Y., Huang, L., Zuo, C., Li, W. & Xing, W. 2020 Novel PVDF-g-NMA copolymer for fabricating the hydrophilic ultrafiltration membrane with good antifouling property. *Industrial & Engineering Chemistry Research* **60** (1), 541–550.

Wu, Q., Tiraferri, A., Wu, H., Xie, W. & Liu, B. 2019 Improving the performance of PVDF/PVDF-g-PEGMA ultrafiltration membranes by partial solvent substitution with Green solvent dimethyl sulfoxide during fabrication. *ACS Omega* **4** (22), 19799–19807.

Xu, R., Feng, Q., He, Y., Yan, F., Chen, L. & Zhao, Y. 2017 Dual functionalized poly(vinylidene fluoride) membrane with acryloylmorpholine and argatroban to improve antifouling and hemocompatibility. *Journal of Biomedical Materials Research Part A* **105** (1), 178–188.

Zhang, X. & Dai, Y. 2019 Recent development of brush polymers via polymerization of poly(ethylene glycol)-based macromonomers. *Polymer Chemistry* **10** (18), 2212–2222.

Zhang, B., Yu, S., Zhu, Y., Shen, Y., Gao, X., Shi, W. & Hwa Tay, J. 2020 Adsorption mechanisms of crude oil onto polytetrafluoroethylene membrane: kinetics and isotherm, and strategies for adsorption fouling control. *Separation and Purification Technology* **235**, 116212.

Zhao, J., Han, H., Wang, Q., Yan, C., Li, D., Yang, J., Feng, X., Yang, N., Zhao, Y. & Chen, L. 2019 Hydrophilic and anti-fouling PVDF blend ultrafiltration membranes using polyacryloylmorpholine-based triblock copolymers as amphiphilic modifiers. *Reactive and Functional Polymers* **139**, 92–101.

Zou, H., Ren, X. & Zhang, J. 2020 Fabrication of a Bi_2O_3 surface-modified polyvinylidene fluoride membrane via an ultraviolet photografting method: improving hydrophilicity and degree of acrylic acid grafting. *Industrial & Engineering Chemistry Research* **59** (14), 6580–6588.

First received 24 June 2021; accepted in revised form 26 August 2021. Available online 7 September 2021

doi: 10.2166/wst.2021.633

An ultrasound/O_3 and UV/O_3 process for atrazine manufacturing wastewater treatment: a multiple scale experimental study

Diya Wen [a,b], Bing Chen[c,*] and Bo Liu[c]

[a] School of Environment, Tsinghua University, Beijing 100084, China
[b] Graduate School at Shenzhen, Tsinghua University, Shenzhen 518055, China
[c] Northern Region Persistent Organic Pollution Control (NRPOP) Laboratory, Faculty of Engineering and Applied Science, Memorial University of Newfoundland, St. John's, NL A1B 3X5, Canada
*Corresponding author. E-mail: bchen@mun.ca

DW, 0000-0001-7081-1448

ABSTRACT

An ultraviolet (UV) and ultrasound (US) enhanced ozonation method were developed to investigate their efficiency on the removal of atrazine and chemical oxygen demand (COD) in authentic atrazine manufacturing wastewater. The bench-scale tests suggested a positive effect of UV and US on the degradation of atrazine within a limited energy range. The pilot-scale flow-through system was further tested by using response surface methodology. The results showed that O_3 and its interaction with UV promoted the degradation of both COD and atrazine while its interaction with US inhibited the removal of COD but promoted the removal of atrazine. The optimal removal rate of atrazine (96.9%) was achieved in the condition of 6.86 W/L UV, 1.96 g/L·h O_3 and 294 W/L US. Chloride ions hindered the atrazine degradation, but the generated free chlorine radicals were still able to react with atrazine. In terms of energy-effectiveness, the configuration of 14.7 W/L UV and 1.96 g/L·h O_3 is the best option, which have the electrical energy per order of 181.6 kWh/m^3 for atrazine and 0.13 kWh/g COD. These method and findings could be helpful in the development of energy-efficient advanced oxidation processes in treating wastewater with high salinity and COD loadings.

Key words: advanced oxidation processes, matrix effect, ozonation, pesticide manufacturing wastewater, ultrasound, UV irradiation

HIGHLIGHTS

- An advanced oxidation process was developed at multiple scales for treating atrazine manufacturing wastewater;
- The excessive energy input of UV and US limited the degradation efficiency of COD;
- Ozone and its interaction with UV promoted the degradation of COD and atrazine;
- Atrazine could be better removed both in bench- and pilot-scale experiements.

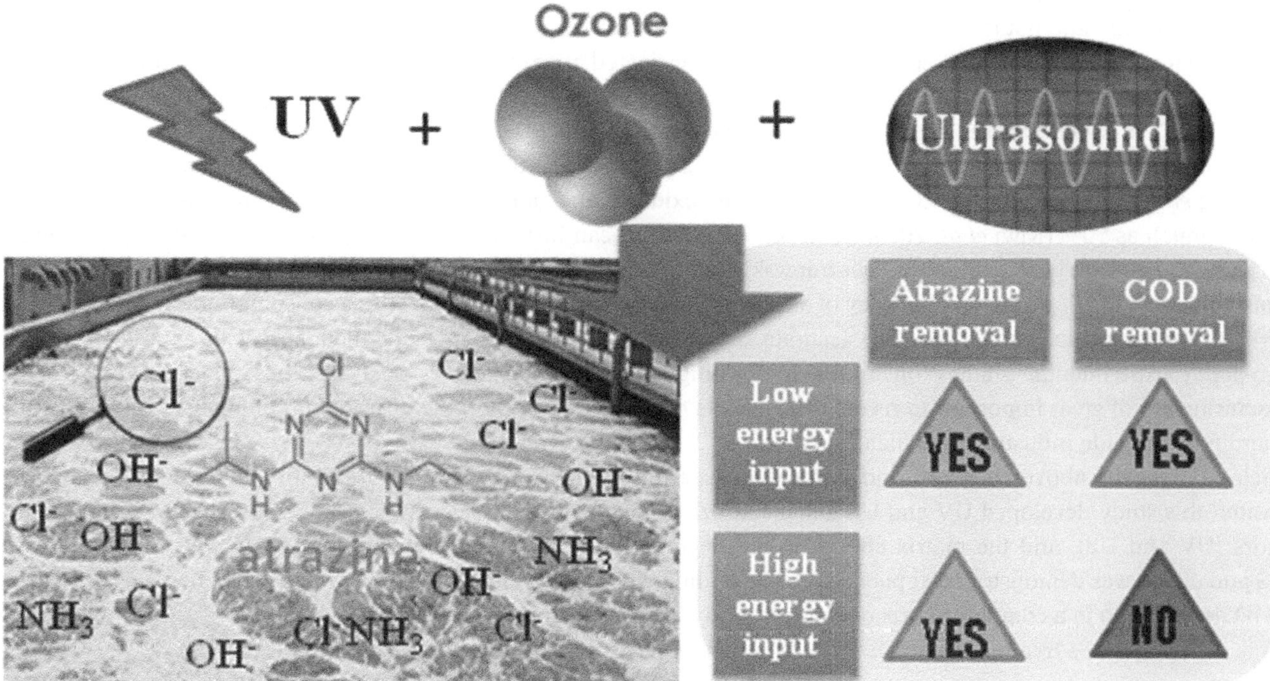

INTRODUCTION

Atrazine (2-chloro-4-(isopropylamino)-6-(ethylamino)-s-triazine), a herbicide commonly used worldwide, has been identified as an endocrine disruptor by the U.S. Environmental Protection Agency (EPA) (Goldman 1994; Dionne *et al.* 2021). It has been listed by as the EPA's Hazardous Waste Hazard Classification System as a persistent and toxic compound (Sonnenschein & Soto 1998; Yang *et al.* 2020). The production of atrazine requires significant amounts of water and trichlorethylene as solvents, leaving a large quantity of wastewater containing organic pollutants such as atrazine, cyanuric chloride, ethylamine and isopropamide (Köck-Schulmeyer *et al.* 2013). This effluent also contains high concentrations of chemical oxygen demand (COD) and ammonia nitrogen (NH₃-N), up to hundreds of thousands mg L^{-1} and over 50 mg L^{-1}, respectively (Janssens *et al.* 1997; Saylor & Kupferle 2019). In addition, high levels of inorganic salt (NaCl), sodium hydroxide (NaOH) and other inorganic substances are commonly observed, which can hinder biochemical treatment processes (Aslan & Şekerdağ 2016; Chen *et al.* 2019a). Conventionally physical/chemical treatment methods followed by a biochemical process are usually used in dealing with atrazine manufacturing wastewater. However, the performance of the physical/chemical treatment often fails to meet the designed efficacy due to the complicated and inconstant composition of the effluent. Further, the biodegradability of atrazine manufacturing wastewater is low owing to its high toxicity and salinity. All of the above lead to substantial challenges in meeting the discharge regulations (Mannina *et al.* 2016; Muñoz Sierra *et al.* 2018; Zhao *et al.* 2020). Therefore, the research and development of more effective treatment methods for atrazine manufacturing wastewater are urgently desired.

Advanced oxidation processes (AOPs) have gained growing interest due to their strength in efficiently breaking down persistent organic pollutants in water, presenting a great potential for treating atrazine manufacturing wastewater. AOPs generate highly reactive species, particularly hydroxyl radicals ('OH), which can effectively remove atrazine and reduce the overall toxicity by generating less toxic compounds (Choi *et al.* 2013; Yang *et al.* 2014). Various sole and combined AOP technologies have been used for treating atrazine contaminated water, including ozonation (Yang *et al.* 2016; Wardenier *et al.* 2019), ultrasound (Collings & Gwan 2010; Liu *et al.* 2017; Pinto *et al.* 2019), peroxide (Li *et al.* 2020a; Song *et al.* 2021), Photo-Fenton and Photo-Fenton like (Khandarkhaeva *et al.* 2017), catalytic oxidation (Zhu *et al.* 2017), and photocatalysis (Yang *et al.* 2020; Zheng *et al.* 2021). Among these technologies, UV/O₃ based treatments are more energy-effective, non-selective of wastewater quality and eco-friendly (Jing *et al.* 2017; Liu *et al.* 2019; Wardenier *et al.* 2019). In addition, the integration of ultrasound (US) not only promotes the transformation of O_3 into ·OH but also generates microbubbles, which provide a large interfacial area for the mass transfer of ozone into liquid phase, thus increase the utility rate of O_3 (Xiong *et al.* 2019).

Although many studies have been reported, only a few investigated atrazine in authentic wastewater with a complex composition of organic and inorganic compounds (Liu *et al.* 2020b). For example, many researches reported that high levels of chloride ion (Cl^-) in water matrices affect AOPs by auxo-action (Xu *et al.* 2013; Monteagudo *et al.* 2016; Mukimin *et al.* 2017; Chen *et al.* 2019b), inhibition (Zhou *et al.* 2017; Zhang *et al.* 2018; Li *et al.* 2020b), and dual functions (Yuan *et al.* 2012b; Luk 2016; Huang *et al.* 2017; Chen *et al.* 2019b). Cl^- consume oxidants and radicals in AOPs, which can reduce the efficacy of AOPs by as much as 50% (Kiwi *et al.* 2000; Chan & Chu 2009). It can further alter the degradation of analytes, resulting in diverse pathways that are hard to monitor (Wiszniowski *et al.* 2003; Li *et al.* 2006; Yuan *et al.* 2011). High COD in wastewater could also diminish the degradation efficiency of analytes by AOPs, owing to the competitive consumption of oxidants by the wastewater substrates (Liu *et al.* 2020a). The accurate analysis of COD was influenced by the matrices (Wayne 1997). Therefore, in-depth experimental studies on AOPs in treating authentic wastewater such as the effluent from atrazine manufacturing are of great important to reveal these effects of water matrices on the degradation of analytes and mineralization, and further guide industrial applications.

To help address the above questions and further improve the performance of AOPs in treating atrazine manufacturing wastewater, this study developed UV and US enhanced ozonation in multiple-scale systems, and investigated the influence of factors (UV and US), and the matrix effects on the degradation efficiency of atrazine and COD. Factorial analysis of UV, US and ozone was demonstrated at pilot scale. The findings should be able to unveil the possible mechanisms of atrazine and COD degradation in a complex matrix of chloride, and nitrogenous compounds, and guide the design and configurations of energy effective AOPs treatment systems for atrazine manufacturing wastewater at large scale.

METHODS

Materials and chemicals

Atrazine-d5 standard and sodium thiosulfate was purchased from Anpel Laboratory Technologies (Shanghai) Inc. Trichloromethane (Thermo Fisher Scientific China) was used for the extraction of atrazine from aqueous samples. Ultrapure water was produced on-site from a Direct-Q 3 UV unit (Millipore, France). Potassium dichromate, sulfuric acid, silver sulfate and mercuric sulfate were purchased from Beijing Chemical Works, China. Atrazine manufacturing wastewater was collected onsite from an agrochemical production plant in China. Detailed location information is not available due to client confidentiality and non-disclosure. The physical-chemical characteristics of the atrazine manufacturing wastewater are summarized in Table 1. The COD was reduced to 6,000–8,000 mg L^{-1} after sand filtration.

The bench-scale tests

As shown in Figure 1, a bench-scale photoreactor has an inner 4 L quartz jar and an outer stainless-steel jacket (Jing *et al.* 2014). The outer jacket has an aluminum lid for heat and light insulation. The inner diameter, height and wall thickness of the quartz jar are 20, 25 and 0.4 cm, respectively. Eight 3.5 W low-pressure UV lamps, emitting exclusively at 254 nm,

Table 1 | Physical-chemical characteristics of the atrazine manufacturing wastewater

Characteristics	Value
COD (mg/L)	14,300
BOD_5 (mg/L)	3,850
TSS (mg/L)	1,890
pH	13–14
Atrazine (mg/L)	5
NH_3-N (mg/L)	50
Chloride (mg/L)	197,500

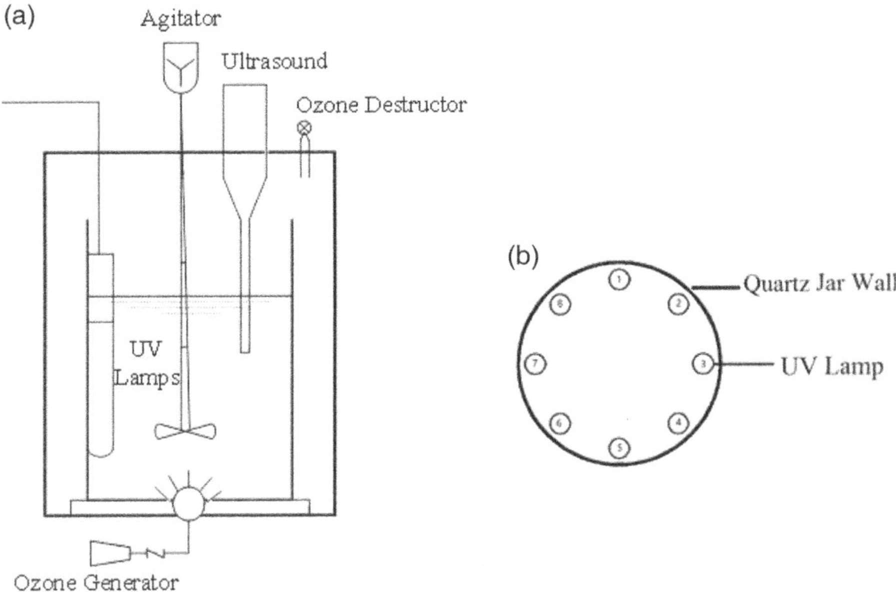

Figure 1 | Bench-scale experiment apparatus (a) schematic diagram, (b) arrangement of eight 3.5 W UV lamps in a vertical view.

are evenly mounted inside the quartz jar near the wall. A 300 W ozone generator with dedicated ozone flow rate monitor are used to produce ozone onsite from ambient air. The ultrasonic system (JY92-IIN, XinZhi Co. Ltd, China) equipped with a probe of which the diameter is 6 mm. The inner quartz jar has a PTFE lid equipped with a stirring rod on which two PTFE six-bladed paddle impellers are mounted to stir the wastewater sample. As to bench-scale experiments, LP-UV lamps were allowed to warm up and reach the stable emission stage for 20 minutes, 2 L water samples were injected into the quartz jar and stirred for uniform mixing. For the UV/O$_3$ test, an ozone generator at a fixed rate of 15 g/h (7.5 g/L·h) was run with the irradiation device simultaneously, and UV power was provided at three different levels of 3.5, 7 and 14 W/L by using 2, 4 and 8 lamps, with light intensities of 2.88, 5.65, 10.93 mw/cm^2, respectively. For the UV/O$_3$/US test, ultrasonic power (20 kHz) was provided at three different levels of 50, 100 and 200 W/L by using an ultrasonic energy converter, and the ultrasonic probe was submerged in the liquid at a level between 15 and 20 mm. During the 180-min reaction period, samples were taken by a peristaltic pump at intervals. In order to eliminate the influence in which water samples remain in tubes, the water samples were taken after pumping for 3 s, transferred into a 20 mL amber vial and immediately quenched by Na$_2$S$_2$O$_3$ to remove residual oxidants.

The pilot-scale tests

A pilot-scale US/UV/O$_3$ system was developed as elaborated previously (Jing *et al.* 2017) (Figure 2). Briefly, a continuous flow-through system contained four reaction columns (cylindrical polycarbonate; 3 feet in height and 10 inches' in internal diameter) with functions of coarse sand filtration, fine sand filtration, ultrasonic ozone treatment and photolytic ozonation, respectively. The low-pressure mercury lamp (UVC-7, LiZhen Co., Ltd, China) was jacked with a quartz filter and the ultraviolet intensity was about 14.5, 15 and 22 mw/cm^2 at UV power of 6.86, 10.8 and 14.7 W/L, respectively. The irradiation and ultrasonic units (US power of 0–1500 W) cooperated with aerating apparatus which can disperse ozone. Peristaltic pumps (YZ1515x, ChuangRui Co., Ltd, China) are used to propel liquids through this system in order to maintain a rather stable flow rate (1.7 L/h). In the pilot-scale tests, water samples passed through filtration and then treated by US/O$_3$ and UV/O$_3$ consequently. The separation of the UV/O$_3$ and US/O$_3$ units was because the synergetic effect of UV and US was marginal (Xu *et al.* 2014). The test procedure was similar to the bench-scale testes, with adjustable factors of UV power (6.86–14.7 W/L), O$_3$ dose (0.98–2.94 g/L·h) and US intensity (0–294 W/L). The detailed experimental procedure was documented in a published paper (Jing *et al.* 2017).

Analytical methods

All laboratory determinations were implemented, following the Standard Methods for Examination of Water and Wastewater. The determination of COD was referred to the standard COD method regulated by the State Environmental

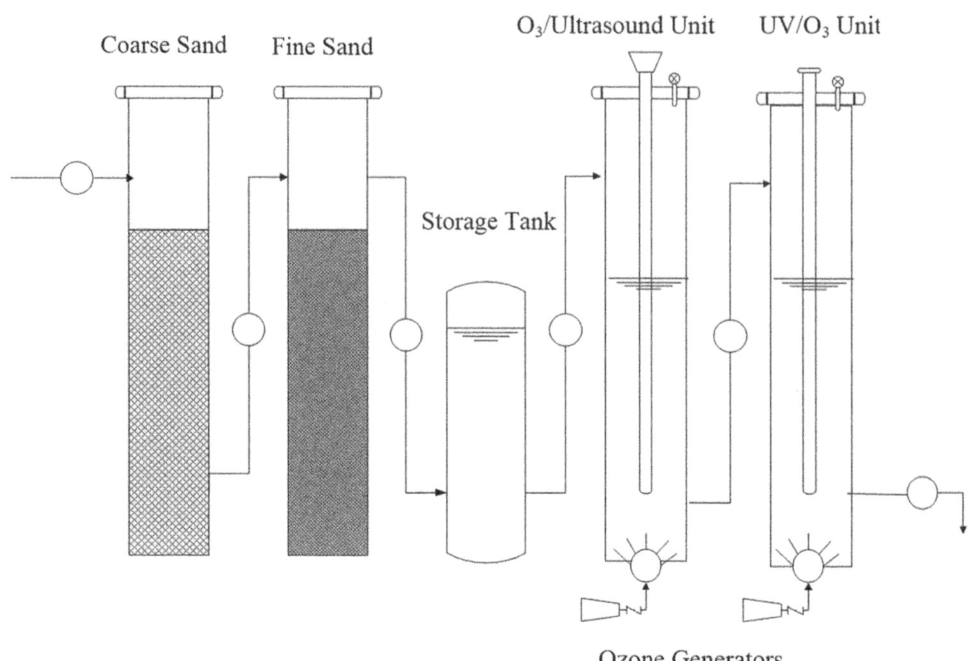

Figure 2 | Schematic diagram of the continuous flow-through system with functions of coarse sand filtration, fine sand filtration, ultrasonic ozone treatment and photolytic ozonation.

Protection Administration, China after dilution due to the high chloride concentration of wastewater samples. The detection limit of the method is 9 mg/L.

Atrazine was extracted by trichloromethane and measured by gas chromatograph-mass spectrometry (GC-MS). An Agilent Technology 7890A GC equipped with an Agilent 5975C MS detector was employed. The gas chromatograph was equipped with an Agilent 7683 autosampler and splitless injector with electronic pressure control. A HP-5MS column (30 m×0.25 mm-×0.25 μm) capillary column was used, with helium as carrier gas at a constant flow of 1 mL min^{-1}. The temperature of the injection port was 250 °C and a 1 μL volume was injected in splitless mode. The GC temperature was ramped from 70 °C, held for 2 min, then to 230 °C at 20 °C·min^{-1} and held for 20 min. The mass spectrometer was operated in EI mode with an ionising energy of 70 eV, ion source temperature at 230 °C, and MS Quad temperature at 150 °C. The mass signal was captured in the SIM mode, scanning from m/z 50 to 500 with a solvent delay at 6 min.

Experimental design and statistical analysis for pilot-scale tests

In order to optimize the UV power (W/L), O$_3$ flow rate (g/L·h) and US power (W/L), and investigate the potential interactions of each factor, central composite design (CCD) and response surface methodology (RSM) were employed for experimental design by Design-Expert® 8.0. According to the parameter levels of the single factor experiment and the treatment capacity of the pilot system, UV power (*A*), O$_3$ flow rate (*B*) and US power (*C*) were amplified to three levels (Table 2), and used −1, 0, 1 on behalf of factors of low, medium, and high values, respectively.

By applying CCD, fewer combinations of the factors are employed to investigate the effects of individual parameters as well as their synergistic interactive effects on the response factors. A general second order polynomial model was selected to explain the behavior of the system (Equation (1))

$$Y = b_0 + \sum_{i=1}^{3} b_i X_i + \sum_{i=1}^{3} b_{ii} X_{ii}^2 + \sum_{i=1}^{3}\sum_{j=1}^{3} b_{ij} X_i X_j \tag{1}$$

where, Y is the response value; b$_0$ is the constant; b$_i$ the linear coefficient; b$_{ii}$ the quadratic coefficient; b$_{ij}$ the interaction coefficient; and X$_i$ dimensionless coded variables (A for UV power, B for O$_3$ flow rate and C for US power). The regression of the

Table 2 | Experimental range and levels of the factor in the CCD design

| | Factor | | |
| | A | B | C |
Level	UV (W/L)	O₃ (g·L/h)	US (W/L)
−1	6.86	0.98	0
0	10.8	1.96	147
+1	14.7	2.94	294

Equation (1) was considered for optimization in order to maximize Y using the numerical optimization program of the same design software.

Response values are COD degradation rate and atrazine removal rate; 20 experiments with three replicates were required for improving response values. The experimental data was further analyzed by assuming a second-order polynomial with linear, quadratic and interaction effects.

RESULTS AND DISCUSSION

Bench-scale experiments

In the UV/O_3 system, when the UV power was increasing from 3.5 to 7 W/L, UV power improved the removal rate of COD, in which the COD removal rates were 57 and 47% at 180 min, respectively (Figure 3(a)). When the UV power exceeded 7 W/L, on the contrary, the COD removal rate did not reduce but increased since the beginning of the reaction. The optimum value of UV power was determined as 7 W/L, in which a removal rate of 94.9% of atrazine was achieved. With the increase of US intensity, the removal rate of COD in the atrazine wastewater was 34.8%, 43.5% and 23.1% at 180 min, respectively (Figure 3(b)). An increase of COD value was observed at the highest US intensity. The best energy-effective value of US power in bench-scale treatment was 100 W/L and 96.5% of atrazine was removed in 10 min under the optimal condition.

Pilot-scale tests and DOE

In order to avoid the growth of COD caused by US, the atrazine manufacturing wastewater was treated in the UV/O_3 and US/O_3 processes consequently. The observed and predicted results of atrazine degradation rate (Y_1, %) and COD removal rate (Y_2, %) are listed in Table S1.

According to statistical model fit summary by Analysis of Variance (ANOVA), a quadratic model (second order polynomial) was selected as the best fitted model. The quadratic effects of the factors on atrazine (Figure 4) and COD

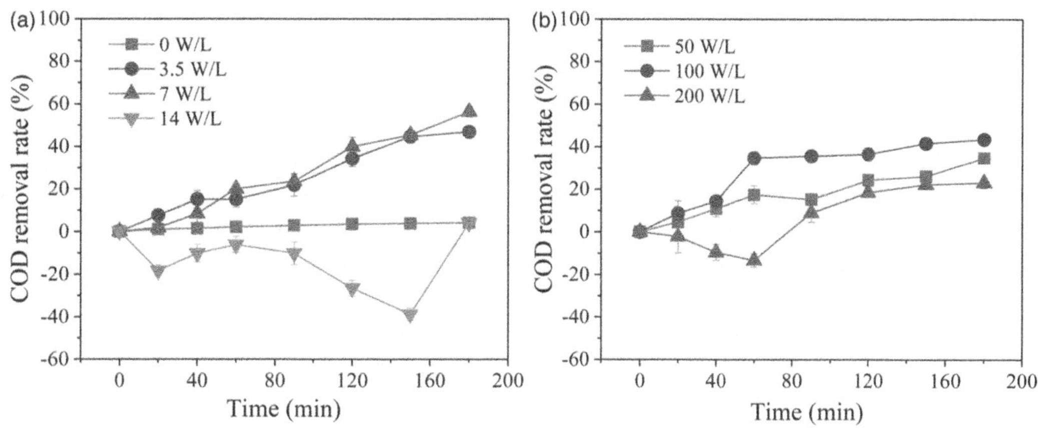

Figure 3 | The COD removal rate of atrazine manufacturing wastewater by (a) the UV/O_3 system and (b) the $UV/O_3/US$ system.

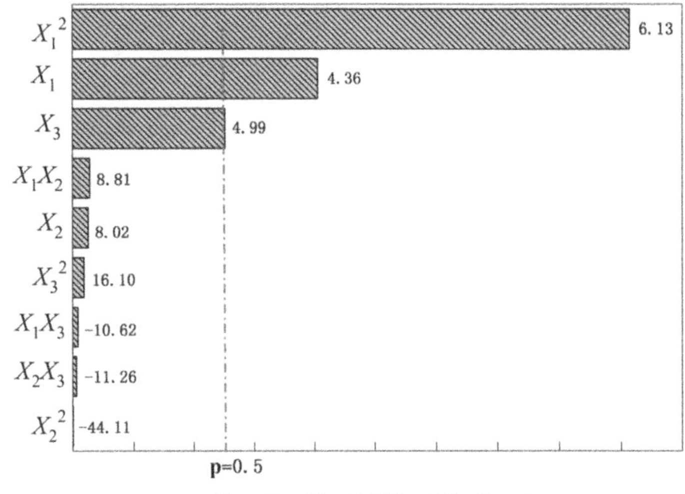

Figure 4 | Standard Pareto chart of the influence of model coefficients on the removal of atrazine.

Table 3 | ANOVA results for the response surface quadratic model on COD removal

Source	SS	DF	MS	F-value	P-value
Model	1,269.5	9	141.1	7.82	0.0017
A-UV	28.0	1	28.02	1.55	0.2409
B-O_3	642.2	1	642.2	35.63	0.0001
C-US	226.0	1	226.2	12.54	0.0054
AB	9.37	1	9.37	0.52	0.4874
AC	0.01	1	0.014	8×10^{-4}	0.9780
BC	65.3	1	65.32	3.62	0.0861
A^2	17.6	1	17.65	0.98	0.3457
B^2	1.62	1	1.62	0.09	0.7708
C^2	113.8	1	113.8	6.31	0.0308
Residual	180.3	10	18.03	–	–
Lack of fit	150.3	5	30.05	5.01	0.0508
Pure error	30.0	5	6.00	–	–
Total	1,449.8	19	–	–	–

(Table 3) were obtained. Figure 4 illustrates that the degradation of atrazine is affected by linearity and quadratic terms of the O_3 flow rate. The possibility of three interaction effects was less than 0.01, which are between UV and O_3 (AB), UV and US (AC), and O_3 flow rate and US (BC), demonstrating there exist interactions among them. Table 3 shows the ANOVA results for the response surface quadratic model on COD removal. The COD removal was sensitive to the linearity term of O_3 flow rate, which agreed with the results of atrazine degradation. However, the possibility of three interaction effects were higher than 0.05, demonstrating no interactions.

The quadratic regression equation for atrazine degradation was established as Equation (2), which shows high R^2 value (0.9498), and acceptable *p*-value (<0.0001). Meanwhile, the regression equation for COD removal rate (Equation (3)) had a low regression ($R^2 = 0.8757$) and an acceptable p-value of 0.0017.

$$\text{Atrazine degradation rate} = 73.0 + 4.4\,A + 8.0\,B + 5.0\,C + 8.81\,AB - 10.6\,AC$$
$$- 11.3\,BC + 6.13A^2 - 44.1B^2 + 16C^2 \tag{2}$$

$$\text{COD removal rate} = 23.7 - 1.67\,A + 8.01\,B - 4.75\,C - 1.08\,AB - 0.04\,AC$$
$$- 2.86\,BC - 2.53\,A^2 + 0.77\,B^2 - 6.43\,C^2 \tag{3}$$

Effect of operational factor on atrazine degradation

The degradation of atrazine in AOPs is mainly through dechlorination and dealkylation, in which the chloride atom on the triazine ring and the alkyl groups on the amines can be attacked and substituted by $^\bullet$OH, forming hydroxyatrazine (OIET) and hydroxydeethyl atrazine (OIAT) and deethyl atrazine (CIAT), deisopropyl atrazine (CIET), and deethyl deisopropyl atrazine (CAAT) (Choi *et al.* 2013; Xu *et al.* 2014; Fan *et al.* 2017). The end product of atrazine in the UV and US process was ammeline (Xu *et al.* 2014). It was suggested that in alkaline solution, indirect oxidation (induced by $^\bullet$OH) dominated (Cuerda-Correa *et al.* 2020). Therefore, the concentration of $^\bullet$OH is of paramount important for atrazine degradation. Equation (8) shows that the most significant factors affecting the degradation of atrazine are O_3 (B), and the interactive effects of UV and O_3 (AB), O_3 and US (BC), and UV and US (AC). The interactive effect of UV and O_3 was positive. In general, the results indicated that more O_3 resulted in a higher concentration of $^\bullet$OH, while higher intensity of UV and US can also produce more $^\bullet$OH, as well as the decomposition of O_3 to $^\bullet$OH. UV and US (AC) showed a strong negative interaction in terms of atrazine degradation. The phenomenon could result from the interference of TSS, and the generation and decomposition of H_2O_2. US can break down the insoluble particles in atrazine wastewater, which would increase the turbidity, increasing the masking effect on UV (Choi *et al.* 2020). In the meantime, US generated a higher level of H_2O_2 (Ziembowicz *et al.* 2017). It thus increased the degradation rate of atrazine. H_2O_2 in alkaline solution can be quickly decomposed in O_2 and water (Nicoll & Smith 1955). UV accelerates the process. It seems that UV with high intensity overaccelerated the decomposition of H_2O_2 generated by US, thus slightly reducing the removal rate of atrazine.

Interactive effect between US and O_3

Bench-scale tests showed an increased atrazine removal to 96.5% by an optimal intensity of US. This aligned with Xu's results, US promotes the generation of $^\bullet$OH by bubble cavitation and attacks the C-Cl position and/or the alkyl side chains of atrazine (Xu *et al.* 2014). Figure 5(b) illustrates the relationship between US power and O_3 flow rate at pilot scale. The US intensity at low and high level had a higher degradation of atrazine. However, the quadratic model (Figure 4) showed that the effect of US was not significant (*P*-value >0.05). This was probably because of the dual effects of US. On one hand, US promoted the decomposition of O_3 and the formation of $^\bullet$OH at low intensity in the vapor phase of cavitated bubbles (Jing *et al.* 2017). On the other hand, the cavitated bubble created a barrier for a biphasic reaction. The interference of the excess ozone could also react with $^\bullet$OH in bubbles, resulting in a recombination of free radicals (Gogate 2008). As the US intensity increased, more barriers were created thus reducing the degradation rate of atrazine. The increase of atrazine degradation at the highest US intensity is probably because of the direct generation of $^\bullet$OH by bubble cavitation. Therefore, it is important to optimize the US and O_3 ratio for the most energy-efficient solutions.

In addition, higher energy US process could result in the release of higher amounts of UV-absorbing compounds, which would reduce the energy-effectiveness of UV significantly. It can be observed in the two configurations of the system with the best atrazine degradation rates. Firstly, an atrazine degradation rate of 96.85% was achieved in the condition of 10.8 W/L UV, 10 g/h O_3 and 294 W/L US. Meanwhile, without US, the atrazine degradation rate can reach 93.93% in the condition of 14.7 W/L UV and 15 g/h O_3.

Interactive effect between UV and O_3

UV irradiation can effectively promote the decomposition of O_3 and the generation of $^\bullet$OH (Beltrán *et al.* 2000; Liu *et al.* 2020a). In bench-scale tests, the degradation of atrazine achieved 94.94% within 10 min, indicating the high efficacy with the optimal UV intensity. However, UV intensity alone was not a significant factor for atrazine degradation at pilot scale.

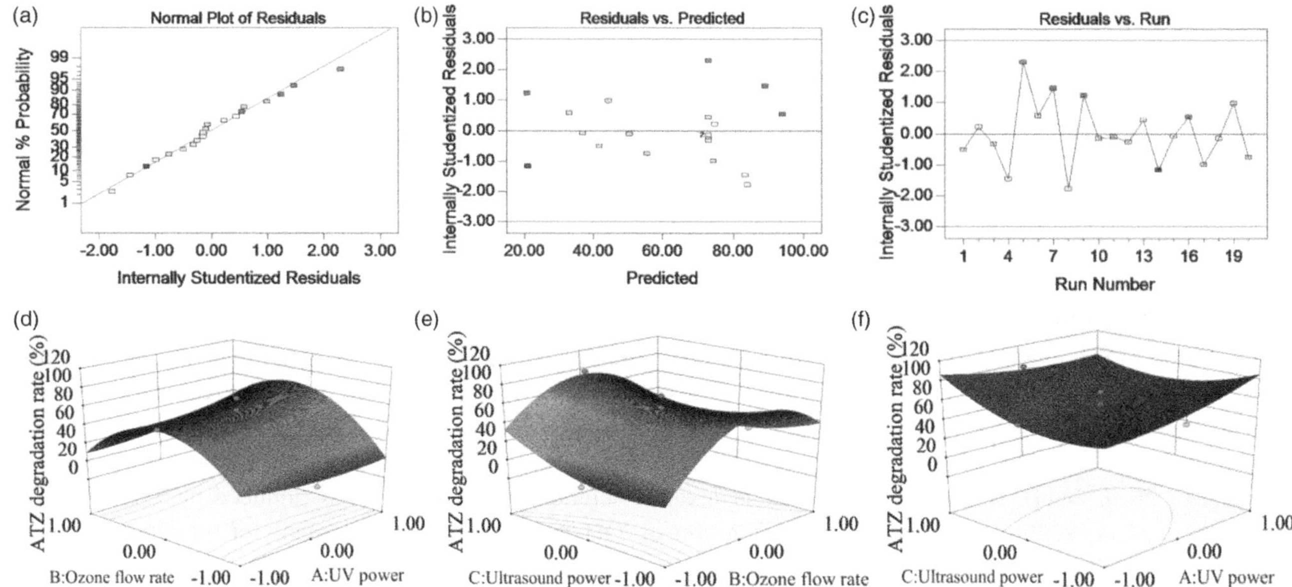

Figure 5 | Response surface analysis results for atrazine removal (a) normality plot of residuals, (b) residuals vs. predicted, (c) residuals vs. run, (d) the interaction effect between O_3 and UV, (e) the interaction effect between O_3 and US, and (f) the interaction effect between UV and US.

Under the condition of low amount of O_3, UV intensity limited enhanced the degradation rate of atrazine, while under the condition of high amount of O_3, UV intensity accelerated the degradation rate of atrazine significantly. It suggested that at low concentration, O_3 was quickly consumed in a high salinity, alkaline and COD environment. The fast decomposition of O_3 in high salinity could limited the effect of UV. When the level of O_3 increased to medium, the acceleration of O_3 decomposition of UV as indicated by the degradation rate is also enhanced. However, higher O_3 dose resulted in a decline of atrazine removal, probably because the excess O_3 might quench $\cdot$OH.

Effect of chloride

High salinity wastewater impacts on the performance of AOPs, as was found in many studies, the summary is shown in Table S2. Organic compounds, including a variety of dye wastewaters and pharmaceutical wastewater, were affected by Cl^-. Some studies suggested that the inhibitory effect of the chloride ion was due to the performance of $\cdot$OH scavengers $(4.3 \times 10^9\,M^{-1}\,s^{-1})$. It competed with organics for the photo-oxidizing species (Guillard *et al.* 2003; Konstantinou & Albanis 2004; Ghodbane & Hamdaoui 2010). In addition, the influence of the chloride ion depended on the pH of the solution. For $\cdot$OH based-AOP, chloride was found to accelerate the degradation of dyes under lower pH conditions, but significant inhibitory effect was found in alkaline condition (Shi-ying *et al.* 2005; Yuan *et al.* 2012a, 2012b). For $\cdot SO_4^-$ based-AOP, normal pH level was found to result in a higher degradation rate of atrazine, because of the inactivation of PMS at acidic pH and precipitation at basic pH (Chan & Chu 2009). Also, the concentration of Cl^- was observed to have a dual effect on AOPs. Studies indicated that lower concentration of Cl^- promoted the discoloration of dyes, which might be attributed to a surface chain-transfer mechanism involving chlorine radicals. But higher Cl^- level reduced the further degradation, which may be caused by the deactivation of the photocatalyst and reduction in photon receiving efficiency (Yuan *et al.* 2012b).

The possible mechanism of reactive oxidative species in high salinity and alkaline conditions is speculated to be as follows: A mass of Cl^- in the atrazine wastewater competed for O_3 with organic matter generated OCl^- (Equation (4)); at higher pH, OCl^- was the predominant species and then was induced by UV to generate chlorine radicals ($\cdot$Cl) (Equation (5)) (Feng *et al.* 2007). Chloride ion also reacted with $\cdot$OH, forming $\cdot OHCl^-$ (Equation (6)). The Cl_2 was produced from $\cdot Cl_2^-$ (Equations (7) and (8)). Cl_2 can be decomposed fast in alkaline solution $(\sim 10^{11}\,M^{-1}\,s^{-1})$ (Knipping & Dabdub 2002), or cleaved into two $\cdot$Cl under UV or high temperature ultrasonic conditions (Equation (10)) thus initiating an additional reaction between $\cdot$Cl and organic matter (Equations (11) and (12)) (Harman 1956; Jing *et al.* 2018).

$$Cl^- + O_3 \rightarrow OCl^- + O_2 \tag{4}$$

$$OCl^- + h\nu \rightarrow {}^\bullet O^- + {}^\bullet Cl \tag{5}$$

$$Cl^- + {}^\bullet OH \rightarrow {}^\bullet OHCl^- \tag{6}$$

$${}^\bullet Cl + Cl^- \rightarrow {}^\bullet Cl_2^- \tag{7}$$

$$2\,{}^\bullet Cl_2^- \rightarrow Cl_2 + 2Cl^- \tag{8}$$

$$Cl_2 + 2OH^- \rightarrow Cl^- + OCl^- + H_2O \tag{9}$$

$$Cl_2 + h\nu \rightarrow {}^\bullet Cl + {}^\bullet Cl \tag{10}$$

$${}^\bullet Cl + RH \rightarrow {}^\bullet R + HCl \tag{11}$$

$${}^\bullet R + Cl_2 \rightarrow RCl + {}^\bullet Cl \tag{12}$$

The presence of chloride slightly affected the degradation of atrazine. Chloride quickly reacted with O_3 and $\cdot$OH, forming less oxidative free chlorine radicals (e.g., $\cdot$Cl, and $\cdot$Cl$_2^-$). But the reaction rate constants of $\cdot$Cl and $\cdot$Cl$_2^-$ with atrazine were 6.87×10^9 M^{-1}s^{-1} and 5×10^4 M^{-1}s^{-1}, respectively (Luo *et al.* 2015; Kong *et al.* 2020). Considering the abundance of free radicals, the overall degradation of atrazine was changed slightly. Nevertheless, the presence of Cl could change the degradation pathways of atrazine. The presence of free chlorine radicals inhibited the dechlorination process, leading to dealkylation as the main process. Atrazine was then gradually decomposed to the intermediates (e.g., CIAT, CIET, CAAT, ammeline) with less toxicity to *Daphnia magna* (Choi *et al.* 2013) and *Vibrio fischeri* (Li & Zhou 2019), which can be further treated by activated sludge (Li *et al.* 2018).

Effect on the COD value

The complete mineralization of one mole of atrazine required 7.5 moles of oxygen (Equation (13)) (Bahena & Martínez 2006). For 5 mg/L atrazine, the theoretical COD value is 5.5 mg/L, which was much lower than the initial COD of the atrazine manufacturing wastewater after sand filtration (6,000–8,000 mg/L). Sometimes atrazine was only oxidized to ammeline without any further change (Xu *et al.* 2014). Therefore, the removal effectiveness of atrazine cannot significantly affect the COD removal.

$$C_8H_{14}ClN_5 + H_2O + 7.5O_2 \rightarrow 8CO_2 + 5NH_3 + HCl \tag{13}$$

The atrazine manufacturing wastewater contained a considerable amount of cyanuric chloride, ethylamine and isopropamide that contributed to the COD value. Equation (3) shows that O_3 dose is the only positive factor in the oxidation of these compounds by providing more oxidants such as O_3 and $\cdot$OH. The increasing trend of COD removal by O_3 indicated that the O_3 dose has not reached the optimal level in the treatment system.

It should be noticed that a high US power resulted in the increase of COD value after treatment. The phenomenon was rarely observed in the clean water studies. For bench-scale tests, the negative effect of COD removal was observed when US power was at 200 W/L. The factorial analysis results of the pilot-scale tests also showed that high US power (294 W/L) had the strongest negative impact to COD removal. The same phenomenon was also found in bench or pilot-scale tests when higher UV power was applied.

Previous research has proved that atrazine-2-hydroxy is the main products of UV photolysis, higher UV power may accelerate the replacement of Cl by OH on atrazine, thus increasing the concentration of free chlorine radicals. In this study, the formation of free chlorine radicals could react with the nitrogenous compounds, forming undesired chloramine, or initiate additional reaction between $\cdot$Cl and organic matter, such as isopropamide in atrazine wastewater (Equations (11) and (12)), which could contribute to the increase of COD value. In addition, the sono-degradation of ATZ was initiated by side chain oxidation; alkylic oxidation products and the dealkylation products prevailed in the early stage and accumulated to their maximum in the middle stage (Xu *et al.* 2014), US can also accelerate the de-chlorination step under ozonation (Bianchi *et al.* 2006). The low frequency US was able to break down the suspended solids in wastewater

(Fetyan & Salem Attia 2020). However, compared with the chemical breakdown, the physical breakdown of the particles was the dominant approach at a low frequency (Mason *et al.* 2011). Therefore, higher power of US may introduce more organic matter in the suspended solids to the aqueous phase without oxidising them, leading to the reduced efficiency of COD removal.

The substrate in the atrazine manufacturing wastewater may interfere with the COD tests in many ways. Although the atrazine manufacturing wastewater has been sand filtrated before oxidation, the soluble residue in the filtrate could still affect the result of COD. The coexist of chloride and nitrogenous compounds (e.g., ammonia and organic amine) in wastewater can produce monochloramine (NH_2Cl) during the acidic $K_2Cr_2O_7$ oxidation, which can severely interfere with the COD value (Wayne 1997). Although the addition of excess silver nitrate during the COD test can remove free chloride ions in the treated effluent, the presence of chloramine in atrazine manufacturing wastewater would affect the COD value. In addition, the COD values of nitrogen-containing heterocycles can be underestimated due to the incomplete oxidation by $K_2Cr_2O_7$. The ratio of N_2 and NH_3 as oxidation products of heterocycles was structure dependent (Chudoba & Dalešický 1973).

The transformation of atrazine, ammeline, triazine (an oxidation product from cyanuric chloride) in the UV/O_3/US system includes deformation, polymerization and oxidation, which could result in both increase and decrease of the COD value; it also explained why the linearity term of UV power and US power had insignificant impact on COD degradation at pilot scale (Figure 5). It was more obvious under the condition of high energy input (e.g., 14 W/L UV, 200 W/L US) but limited oxidant level in the bench-scale experiments, in which the deformation of nitrogen-containing heterocycles by UV was the major reaction instead of oxidation.

Energy consumption

The performance evaluation of combined AOPs should also consider the energy consumption for a certain target removal (Tijani *et al.* 2014; Miklos *et al.* 2018; Jiménez *et al.* 2019). Many studies have use electrical energy per order of target compound destruction (EEO) to determine the energy-effectiveness (Choi *et al.* 2019; Jiménez *et al.* 2019). The EEO (kWh/m^3) can be calculated using the equation below:

$$EEO = \frac{P \times t \times 1000}{V \times \log\left(\frac{C_i}{C_f}\right)} \tag{14}$$

where P is the total energy input of the processes (kW), V is the volume (L) of the water treated in time t (h), and C_i and C_f are the initial and final concentrations of targets (mg/L).

The EEO of atrazine under different AOPs combinations are shown in Table 4. The processes with US generally consumed more energy than other processes due to the high energy input of US. However, it did not show a corresponding enhancement of atrazine degradation. Most of the US-enhanced processes have EEO values higher than 1,000 kWh/m^3. The process with the highest atrazine removal (Run 16, 96.9%) consumed 684.29 kWh/m^3, while the process with the second highest atrazine removal (Run 7, 93.9%) only had an EEO value of 181.6 kWh/m^3. It is suggested that US was not an energy-effective addition for ATZ degradation.

On the other hand, the wastewater matrix can strongly interfere with the removal efficiency of target compounds. The competitive consumption of wastewater substrates reduces the overall oxidation rates (Liu *et al.* 2020a), which on other hand require much higher energy to reach the same removal rates of target compounds in the distilled water system. Miklos *et al.* (2018) summarized the EEOs of O_3 based AOPs. They pointed out that the median value of EEO in these processes was <1 kWh/m^3. However, the EEO value for treating the real wastewater using ozonation (188.01 kWh/m^3) can be found at two orders of magnitude higher (Jiménez *et al.* 2019). EEO omits the effects of the water matrix thus is not suitable to evaluate the energy-effectiveness of different treatment processes for the cleanup of water with different matrices.

For COD, the electron energy consumption per COD removal would be a better indicator. Table 4 shows the energy consumption rates of different runs. It is indicated that US has significantly increased the energy consumption rates. The lowest rate of 0.1 kWh/g can be found in Run 1, which was with an energy input of 6.86 W/L from UV and 2.94 g/h from O_3. However, its EEO value for atrazine was 888.26 kWh/m^3, which was much higher that Run 7. In the consideration of both atrazine and COD removal, Run 7 is the most energy-effective treatment option for atrazine manufacturing wastewater.

Table 4 | Electric energy consumption of different AOPs combinations

Run No.	UV W/L	O_3 g/L·h	US W/L	Energy consumption Atrazine kWh/m³	COD kWh/g
1	6.86	2.94	0	888.26	0.1
2	6.86	1.96	147	937.59	0.37
3	10.8	1.96	147	1,102.72	0.47
4	14.7	1.96	147	969.98	0.84
5	10.8	1.96	147	631.91	0.55
6	14.7	0.98	0	552.2	0.16
7	14.7	2.94	0	181.6	0.13
8	10.8	1.96	0	248.43	0.12
9	6.86	0.98	0	644.55	0.12
10	10.8	1.96	147	1,069.34	0.41
11	6.86	2.94	294	3,585.73	1.02
12	10.8	1.96	147	1,090.49	0.46
13	10.8	1.96	147	953.84	0.38
14	10.8	0.98	147	7,542.43	0.46
15	10.8	2.94	147	3,284.31	0.24
16	10.8	1.96	294	684.29	1.11
17	6.86	0.98	294	1,774.09	4.47
18	10.8	1.96	147	1,069.34	0.52
19	14.7	0.98	294	3,520.91	2.11
20	14.7	2.94	294	3,363.63	1.55

CONCLUSION

Conventional physical/chemical/biological treatments of atrazine manufacturing wastewater containing high levels of chloride, COD and alkaline that usually failed to meet the discharge regulations. The efficiency and the energy-effectiveness of the UV/O_3/US processes were investigated in treating the atrazine manufacturing wastewater through the bench- and pilot-scale systems. Bench-scale tests showed that at the levels of 7 W/L of UV and 100 W/L of US, the integration of UV and US in ozonation increased the atrazine degradation and COD removal. A 96.5% removal of atrazine was achieved in the best condition. However, the excessive energy inputs (e.g., UV≥14 W/L and US ≥200 W/L) could increase the COD value of the effluent. In the pilot-scale system, the significant factors were O_3, and the interactive effects of UV and O_3, O_3 and US, and UV and US. ·OH induced oxidation is the dominated process for atrazine degradation in the alkaline condition. High US intensity might enhance atrazine degradation by directly generating ·OH from cavitation. However, high US intensity could also promote the UV absorbance of the background water matrix, thus reducing the energy-effectiveness of UV. In terms of matrix effect, high salinity slightly reduced the removal of atrazine. By calculating the energy consumption rates of different configurations in the pilot-scale system, the most energy-effective option for atrazine and COD removal is 14.7 W/L UV and 7.5 g/L·h O_3, which have the removal rates of 93.9% and 23.7% for atrazine and COD, respectively.

In summary, the UV/O_3/US process could effectively degrade atrazine in the manufacturing wastewater, while US is less energy effective. The COD value of the effluent during and after treatment was affected by O_3. The bidirectional effects of energy inputs (e.g., UV, US) in the presence of insoluble particles, chloride, and nitrogenous compounds were observed in the removal of COD in the wastewater.

Although the system showed a promising performance in the removal of atrazine from the wastewater, the removal of COD can be further improved by optimizing the operational factors, quantifying the interference caused by the wastewater matrices, and reducing the 'noise' to the COD analysis. In addition, the primarily treated wastewater is usually collected and further treated in industrial wastewater treatment plants, which is dominated by the biochemical treatment systems (e.g., activated sludge). Therefore, further investigation on the end products and other composition in treated atrazine wastewater, as well as its toxicity is necessary.

ACKNOWLEDGEMENT

Special thanks go to Natural Sciences and Engineering Research Council of Canada (NSERC) and Canada Foundation for Innovation (CFI) for supporting this research.

DATA AVAILABILITY STATEMENT

All relevant data are included in the paper or its Supplementary Information.

REFERENCES

Aslan, S. & Şekerdağ, N. 2016 Salt inhibition on anaerobic treatment of high salinity wastewater by upflow anaerobic sludge blanket (UASB) reactor. *Desalination and Water Treatment* **57** (28), 12998–13004.

Bahena, C. L. & Martínez, S. S. 2006 Photodegradation of chlorbromuron, atrazine, and alachlor in aqueous systems under solar irradiation. *International Journal of Photoenergy* **2006**, 1–6.

Beltrán, F. J., González, M., Acedo, B. & Rivas, F. J. 2000 Kinetic modelling of aqueous atrazine ozonation processes in a continuous flow bubble contactor. *Journal of Hazardous Materials* **80** (1), 189–206.

Bianchi, C. L., Pirola, C., Ragaini, V. & Selli, E. 2006 Mechanism and efficiency of atrazine degradation under combined oxidation processes. *Applied Catalysis B: Environmental* **64** (1–2), 131–138.

Chan, K. H. & Chu, W. 2009 Degradation of atrazine by cobalt-mediated activation of peroxymonosulfate: different cobalt counteranions in homogenous process and cobalt oxide catalysts in photolytic heterogeneous process. *Water Research* **43** (9), 2513–2521.

Chen, L., Hu, Q., Zhang, X., Chen, Z., Wang, Y. & Liu, S. 2019a Effects of salinity on the biological performance of anaerobic membrane bioreactor. *Journal of Environmental Management* **238**, 263–273.

Chen, L. W., Zuo, X., Yang, S. J., Cai, T. M. & Ding, D. H. 2019b Rational design and synthesis of hollow Co_3O_4@Fe_2O_3 core-shell nanostructure for the catalytic degradation of norfloxacin by coupling with peroxymonosulfate. *Chemical Engineering Journal* **359**, 373–384.

Choi, J., Cui, M., Lee, Y., Ma, J., Kim, J., Son, Y. & Khim, J. 2019 Hybrid reactor based on hydrodynamic cavitation, ozonation, and persulfate oxidation for oxalic acid decomposition during rare-earth extraction processes. *Ultrasonics Sonochemistry* **52**, 326–335.

Choi, H. J., Kim, D. & Lee, T. J. 2013 Photochemical degradation of atrazine in UV and UV/H_2O_2 process: pathways and toxic effects of products. *Journal of Environmental Science and Health Part B Pesticides Food Contaminants and Agricultural Wastes* **48** (11), 927–934.

Choi, Y., Lee, D., Hong, S., Khan, S., Darya, B., Lee, J. Y. & Cho, S. H. 2020 Investigation of the synergistic effect of sonolysis and photocatalysis of titanium dioxide for organic dye degradation. *Catalysts* **10** (5), 500.

Chudoba, J. & Dalešický, J. 1973 Chemical oxygen demand of some nitrogenous heterocyclic compounds. *Water Research* **7** (5), 663–668.

Collings, A. F. & Gwan, P. B. 2010 Ultrasonic destruction of pesticide contaminants in slurries. *Ultrasonics Sonochemistry* **17** (1), 1–3.

Cuerda-Correa, E. M., Alexandre-Franco, M. F. & Fernández-González, C. 2020 Advanced oxidation processes for the removal of antibiotics from water. An overview. *Water* **12** (1), 102.

Dionne, E., Hanson, M. L., Anderson, J. C. & Brain, R. A. 2021 Chronic toxicity of technical atrazine to the fathead minnow (Pimephales promelas) during a full life-cycle exposure and an evaluation of the consistency of responses. *Science of the Total Environment* **755**, 142589.

Fan, Y., Ji, Y., Zheng, G., Lu, J., Kong, D., Yin, X. & Zhou, Q. 2017 Degradation of atrazine in heterogeneous Co_3O_4 activated peroxymonosulfate oxidation process: kinetics, mechanisms, and reaction pathways. *Chemical Engineering Journal* **330**, 831–839.

Feng, Y., Smith, D. W. & Bolton, J. R. 2007 Photolysis of aqueous free chlorine species (HOCl and OCl$^-$) with 254 nm ultraviolet light. *Journal of Environmental Engineering and Science* **6** (3), 277–284.

Fetyan, N. A. H. & Salem Attia, T. M. 2020 Water purification using ultrasound waves: application and challenges. *Arab Journal of Basic and Applied Sciences* **27** (1), 194–207.

Ghodbane, H. & Hamdaoui, O. 2010 Decolorization of antraquinonic dye, C.I. Acid Blue 25, in aqueous solution by direct UV irradiation, UV/H_2O_2 and UV/Fe(II) processes. *Chemical Engineering Journal* **160** (1), 226–231.

Gogate, P. R. 2008 Cavitational reactors for process intensification of chemical processing applications: a critical review. *Chemical Engineering and Processing: Process Intensification* **47** (4), 515–527.

Goldman, L. 1994 *Atrazine, Simazine and Cyanazine: Notice of Initiation of Special Review*. Federal Register, EPA, Washington, pp. 60412–60443.

Guillard, C., Lachheb, H., Houas, A., Ksibi, M., Elaloui, E. & Herrmann, J.-M. 2003 Influence of chemical structure of dyes, of pH and of inorganic salts on their photocatalytic degradation by TiO_2 comparison of the efficiency of powder and supported TiO_2. *Journal of Photochemistry and Photobiology A: Chemistry* **158** (1), 27–36.

Harman, D. 1956 Aging: a theory based on free radical and radiation chemistry. *Journal of Gerontology* **11** (3), 298–300.

Huang, T., Chen, J., Wang, Z., Guo, X. & Crittenden, J. C. 2017 Excellent performance of cobalt-impregnated activated carbon in peroxymonosulfate activation for acid orange 7 oxidation. *Environmental Science and Pollution Research* **24** (10), 9651–9661.

Janssens, I., Tanghe, T. & Verstraete, W. 1997 Micropollutants: a bottleneck in sustainable wastewater treatment. *Water Science and Technology* **35** (10), 13–26.

Jiménez, S., Andreozzi, M., Micó, M. M., Álvarez, M. G. & Contreras, S. 2019 Produced water treatment by advanced oxidation processes. *Science of the Total Environment* **666**, 12–21.

Jing, L., Chen, B., Zhang, B., Zheng, J. & Liu, B. 2014 Naphthalene degradation in seawater by UV irradiation: the effects of fluence rate, salinity, temperature and initial concentration. *Marine Pollution Bulletin* **81** (1), 149.

Jing, L., Chen, B., Wen, D., Zheng, J. & Zhang, B. 2017 Pilot-scale treatment of atrazine production wastewater by UV/O_3/ultrasound: factor effects and system optimization. *Journal of Environmental Management* **203** (Pt 1), 182–190.

Jing, L., Chen, B., Wen, D., Zheng, J. & Zhang, B. 2018 The removal of COD and NH_3-N from atrazine production wastewater treatment using UV/O_3: experimental investigation and kinetic modeling. *Environmental Science and Pollution Research International* **25** (3), 2691–2701.

Khandarkhaeva, M., Batoeva, A., Aseev, D., Sizykh, M. & Tsydenova, O. 2017 Oxidation of atrazine in aqueous media by solar- enhanced Fenton-like process involving persulfate and ferrous ion. *Ecotoxicology and Environment Safety* **137**, 35–41.

Kiwi, J., Lopez, A. & Nadtochenko, V. 2000 Mechanism and kinetics of the OH-radical intervention during Fenton oxidation in the presence of a significant amount of radical scavenger (Cl^-). *Environmental Science & Technology* **34** (11), 2162–2168.

Knipping, E. M. & Dabdub, D. 2002 Modeling Cl_2 formation from aqueous NaCl particles: evidence for interfacial reactions and importance of Cl_2 decomposition in alkaline solution. *Journal of Geophysical Research: Atmospheres* **107** (D18), ACH 8-1–ACH 8-30.

Köck-Schulmeyer, M., Villagrasa, M., de Alda, M. L., Céspedes-Sánchez, R., Ventura, F. & Barceló, D. 2013 Occurrence and behavior of pesticides in wastewater treatment plants and their environmental impact. *Science of the Total Environment* **458**, 466–476.

Kong, X., Wang, L., Wu, Z., Zeng, F., Sun, H., Guo, K., Hua, Z. & Fang, J. 2020 Solar irradiation combined with chlorine can detoxify herbicides. *Water Research* **177**, 115784.

Konstantinou, I. K. & Albanis, T. A. 2004 TiO_2-assisted photocatalytic degradation of azo dyes in aqueous solution: kinetic and mechanistic investigations. *Applied Catalysis B: Environmental* **49** (1), 1–14.

Li, H. & Zhou, B. 2019 Degradation of atrazine by catalytic ozonation in the presence of iron scraps: performance, transformation pathway, and acute toxicity. *Journal of Environmental Science and Health, Part B* **54** (5), 432–440.

Li, G., An, T., Chen, J., Sheng, G., Fu, J., Chen, F., Zhang, S. & Zhao, H. 2006 Photoelectrocatalytic decontamination of oilfield produced wastewater containing refractory organic pollutants in the presence of high concentration of chloride ions. *Journal of Hazardous Materials* **138** (2), 392.

Li, J., Xu, M., Yao, G. & Lai, B. 2018 Enhancement of the degradation of atrazine through $CoFe_2O_4$ activated peroxymonosulfate (PMS) process: kinetic, degradation intermediates, and toxicity evaluation. *Chemical Engineering Journal* **348**, 1012–1024.

Li, G., Wang, C., Yan, Y. P., Yan, X. R., Li, W. T., Feng, X. H., Li, J. S., Xiang, Q. J., Tan, W. F., Liu, F. & Yin, H. 2020a Highly enhanced degradation of organic pollutants in hematite/sulfite/photo system. *Chemical Engineering Journal* **386**, 12.

Li, Z., Sun, Y., Yang, Y., Han, Y., Wang, T., Chen, J. & Tsang, D. C. W. 2020b Biochar-supported nanoscale zero-valent iron as an efficient catalyst for organic degradation in groundwater. *Journal of Hazardous Materials* **383**, 121240.

Liu, C., Sun, Y., Wang, D., Sun, Z., Chen, M., Zhou, Z. & Chen, W. 2017 Performance and mechanism of low-frequency ultrasound to regenerate the biological activated carbon. *Ultrasonics Sonochemistry* **34**, 142–153.

Liu, Z., Hosseinzadeh, S., Wardenier, N., Verheust, Y., Chys, M. & Hulle, S. V. 2019 Combining ozone with UV and H_2O_2 for the degradation of micropollutants from different origins: lab-scale analysis and optimization. *Environmental Technology* **40** (28), 3773–3782.

Liu, B., Chen, B., Zhang, B., Song, X., Zeng, G. & Lee, K. 2020a Photocatalytic ozonation of offshore produced water by TiO_2 nanotube arrays coupled with UV-LED irradiation. *Journal of Hazardous Materials* **402**, 123456.

Liu, Z., Yang, Y., Shao, C., Ji, Z., Wang, Q., Wang, S., Guo, Y., Demeestere, K. & Hulle, S. V. 2020b Ozonation of trace organic compounds in different municipal and industrial wastewaters: kinetic-based prediction of removal efficiency and ozone dose requirements. *Chemical Engineering Journal* **387**, 123405.

Luk, M. K. 2016 *Photocatalytic Degradation and Chlorination of Azo Dye in Saline Wastewater.* UTAR, Perak, Malaysia.

Luo, C., Ma, J., Jiang, J., Liu, Y., Song, Y., Yang, Y., Guan, Y. & Wu, D. 2015 Simulation and comparative study on the oxidation kinetics of atrazine by UV/H_2O_2, UV/HSO_5^- and $UV/S_2O_8^{2-}$. *Water Research* **80**, 99–108.

Mannina, G., Capodici, M., Cosenza, A., Di Trapani, D. & Viviani, G. 2016 Sequential batch membrane bio-reactor for wastewater treatment: the effect of increased salinity. *Bioresource Technology* **209**, 205–212.

Mason, T. J., Cobley, A. J., Graves, J. E. & Morgan, D. 2011 New evidence for the inverse dependence of mechanical and chemical effects on the frequency of ultrasound. *Ultrasonics Sonochemistry* **18** (1), 226–230.

Miklos, D. B., Remy, C., Jekel, M., Linden, K. G., Drewes, J. E. & Hübner, U. 2018 Evaluation of advanced oxidation processes for water and wastewater treatment – a critical review. *Water Research* **139**, 118–131.

Monteagudo, J. M., Duran, A., Latorre, J. & Exposito, A. J. 2016 Application of activated persulfate for removal of intermediates from antipyrine wastewater degradation refractory towards hydroxyl radical. *Journal of Hazardous Materials* **306**, 77–86.

Mukimin, A., Zen, N., Purwanto, A., Wicaksono, K. A., Vistanty, H. & Alfauzi, A. S. 2017 Application of a full-scale electrocatalytic reactor as real batik printing wastewater treatment by indirect oxidation process. *Journal of Environmental Chemical Engineering* **5** (5), 5222–5232.

Muñoz Sierra, J. D., Oosterkamp, M. J., Wang, W., Spanjers, H. & van Lier, J. B. 2018 Impact of long-term salinity exposure in anaerobic membrane bioreactors treating phenolic wastewater: performance robustness and endured microbial community. *Water Research* **141**, 172–184.

Nicoll, W. D. & Smith, A. F. 1955 Stability of dilute alkaline solutions of hydrogen peroxide. *Industrial & Engineering Chemistry* **47** (12), 2548–2554.

Pinto, C. F., Antonelli, R., de Araújo, K. S., Fornazari, A. L. d. T., Fernandes, D. M., Granato, A. C., Azevedo, E. B. & Malpass, G. R. P. 2019 Experimental-design-guided approach for the removal of atrazine by sono-electrochemical-UV-chlorine techniques. *Environmental Technology* **40** (4), 430–440.

Saylor, G. L. & Kupferle, M. J. 2019 The impact of chloride or bromide ions on the advanced oxidation of atrazine by combined electrolysis and ozonation. *Journal of Environmental Chemical Engineering* **7** (3), 103105.

Shi-ying, Y., Ying-xu, C., Li-ping, L. & Xiao-na, W. 2005 Involvement of chloride anion in photocatalytic process. *Journal of Environmental Sciences* **17** (5), 761–765.

Song, W., Li, J., Fu, C., Wang, Z., Zhou, Y., Zhang, X., Yang, J., Wang, K., Liu, Y. & Song, Q. 2021 Establishment of sulfate radical advanced oxidation process based on Fe^{2+}/O_2/dithionite for organic contaminants degradation. *Chemical Engineering Journal* **410**, 128204.

Sonnenschein, C. & Soto, A. M. 1998 An updated review of environmental estrogen and androgen mimics and antagonists. *Journal of Steroid Biochemistry and Molecular Biology* **65** (1–6), 143–150.

Tijani, J. O., Fatoba, O. O., Madzivire, G. & Petrik, L. F. 2014 A review of combined advanced oxidation technologies for the removal of organic pollutants from water. *Water, Air, & Soil Pollution* **225** (9), 1–30.

Wardenier, N., Liu, Z., Nikiforov, A., Van Hulle, S. W. H. & Leys, C. 2019 Micropollutant elimination by O_3, UV and plasma-based AOPs: an evaluation of treatment and energy costs. *Chemosphere* **234**, 715–724.

Wayne, B. 1997 *The Science of Chemical Oxygen Demand Technical Information Series*. Hach Company, Ames, IA, USA, p. 22.

Wiszniowski, J., Robert, D., Surmacz-Gorska, J., Miksch, K. & Weber, J.-V. 2003 Photocatalytic mineralization of humic acids with TiO_2: effect of pH, sulfate and chloride anions. *International Journal of Photoenergy* **5** (2), 69.

Xiong, X., Wang, B., Zhu, W., Tian, K. & Zhang, H. 2019 A review on ultrasonic catalytic microbubbles ozonation processes: properties, hydroxyl radicals generation pathway and potential in application. *Catalysts* **9** (1), 10.

Xu, L., Yuan, R., Guo, Y., Xiao, D., Cao, Y., Wang, Z. & Liu, J. 2013 Sulfate radical-induced degradation of 2,4,6-trichlorophenol: a de novo formation of chlorinated compounds. *Chemical Engineering Journal* **217**, 169–173.

Xu, L. J., Chu, W. & Graham, N. 2014 Atrazine degradation using chemical-free process of USUV: analysis of the micro-heterogeneous environments and the degradation mechanisms. *Journal of Hazardous Materials* **275**, 166–174.

Yang, Y., Cao, H., Peng, P. & Bo, H. 2014 Degradation and transformation of atrazine under catalyzed ozonation process with TiO_2 as catalyst. *Journal of Hazardous Materials* **279**, 444–451.

Yang, J., Li, J., Dong, W., Ma, J., Cao, J., Li, T., Li, J., Gu, J. & Liu, P. 2016 Study on enhanced degradation of atrazine by ozonation in the presence of hydroxylamine. *Journal of Hazardous Materials* **316**, 110–121.

Yang, L., Yao, G. & Huang, S. 2020 Enhanced degradation of atrazine in water by VUV/UV/Fe process: role of the in situ generated H_2O_2. *Chemical Engineering Journal* **388**, 124302.

Yuan, R., Ramjaun, S. N., Wang, Z. & Liu, J. 2011 Effects of chloride ion on degradation of Acid Orange 7 by sulfate radical-based advanced oxidation process: implications for formation of chlorinated aromatic compounds. *Journal of Hazardous Materials* **196**, 173–179.

Yuan, R., Ramjaun, S. N., Wang, Z. & Liu, J. 2012a Concentration profiles of chlorine radicals and their significances in ˙OH-induced dye degradation: kinetic modeling and reaction pathways. *Chemical Engineering Journal* **209**, 38–45.

Yuan, R., Ramjaun, S. N., Wang, Z. & Liu, J. 2012b Photocatalytic degradation and chlorination of azo dye in saline wastewater: kinetics and AOX formation. *Chemical Engineering Journal* **192**, 171–178.

Zhang, W., Zhou, S., Sun, J., Meng, X., Luo, J., Zhou, D. & Crittenden, J. 2018 Impact of chloride ions on UV/H_2O_2 and UV/Persulfate advanced oxidation processes. *Environmental Science and Technology* **52** (13), 7380–7389.

Zhao, Y., Zhuang, X., Ahmad, S., Sung, S. & Ni, S. Q. 2020 Biotreatment of high-salinity wastewater: current methods and future directions. *World Journal of Microbiology and Biotechnology* **36** (3), 37.

Zheng, Z., Zhang, K., Toe, C. Y., Amal, R., Zhang, X., McCarthy, D. T. & Deletic, A. 2021 Stormwater herbicides removal with a solar-driven advanced oxidation process: a feasibility investigation. *Water Research* **190**, 116783.

Zhou, L., Ferronato, C., Chovelon, J.-M., Sleiman, M. & Richard, C. 2017 Investigations of diatrizoate degradation by photo-activated persulfate. *Chemical Engineering Journal* **311**, 28–36.

Zhu, S., Dong, B., Yu, Y., Bu, L., Deng, J. & Zhou, S. 2017 Heterogeneous catalysis of ozone using ordered mesoporous Fe_3O_4 for degradation of atrazine. *Chemical Engineering Journal* **328**, 527–535.

Ziembowicz, S., Kida, M. & Koszelnik, P. 2017 Sonochemical formation of hydrogen peroxide. *Proceedings* **2** (5), 188.

First received 17 June 2021; accepted in revised form 13 December 2021. Available online 23 December 2021

doi: 10.2166/wst.2022.178

Enhancement of E-Peroxone process with waste-tire carbon composite cathode for tinidazole degradation

Anlin Xu, Wanqun Liu, Leping Chu, Yunhai Zhang, Yide He and Yongjun Zhang [*]

School of Environmental Engineering and Science, Nanjing Tech University, Nanjing 211816, China
*Corresponding author. E-mail: y.zhang@njtech.edu.cn

YZ, 0000-0001-5562-9090

ABSTRACT

The cathode is the key component in the electro-peroxone process (E-Peroxone), which is popularly constructed with carbon materials. This study developed an innovative method to fabricate a cathode with waste-tire carbon (WTC) whose performance was evaluated for the degradation of tinidazole (TNZ), an antibiotic frequently detected in water. It was found that the addition of WTC in the cathode can significantly promote the yield of H_2O_2 and the current efficiency: around 2.7 times that of commercial carbon black at the same loading. The critical influencing factors were studied, including the current density, ozone concentration, initial pH value, chlorine ions and initial TNZ concentration. The scavenger tests demonstrated the possible involvement of •OH and $O_2^{•-}$. Some transformation products of TNZ were identified with UPLC-MS and the degradation pathway was proposed accordingly. These results demonstrated the potential of WTC for developing E-Peroxone cathodes.

Key words: AOP, emerging organic contaminants, ozonation, perchlorate, recalcitrant

HIGHLIGHTS

- The cathode with WTC generated more H_2O_2 than a commercial carbon black cathode.
- Reaction conditions including current density, O_3 dose, pH values, initial TNZ concentration and electrolyte were studied.
- The main reactive oxygen species were probably •OH and $O_2^{•-}$.
- Transformation products were identified and a degradation pathway was proposed.

1. INTRODUCTION

Advanced oxidation processes (AOPs) are highly effective to degrade the organic contaminants in the aquatic environment due to the generation of reactive oxidation species, such as $SO_4^{•-}$, •OH, $O_2^{•-}$ and $^1O_2$ (Turkay *et al.* 2017). Among them, hydrogen peroxide plus ozone (H_2O_2/O_3), also called peroxone, has certain advantages like easy scale-up, simple operation, and high effectiveness, and thus attracts a lot of research interest. However, due to its hazardous properties such as oxidation, skin irritation, corrosion, etc., the transport and storage of H_2O_2 should be carefully handled during the application of peroxone (Brillas 2022). To overcome the problems, an electro-peroxone (E-Peroxone) process has been developed by combining the electrogeneration of H_2O_2 and ozonation in a reactor, in which O_2 is transformed to H_2O_2 at the cathode surface via a two-electron reduction reaction (Equation (1)) when a mixture of O_2 and O_3 gas, released from an O_3 generator, is bubbled into the reactor (Mao *et al.* 2018). Then, the generation of •OH will be promoted with the same mechanisms with conventional peroxone (simplified in Equation (2)):

$$O_2 + 2H^+ + 2e^- \rightarrow H_2O_2 \tag{1}$$

$$H_2O_2 + O_3 \rightarrow •OH + •O_2^- + H^+ + O_2 \tag{2}$$

Besides the enhanced generation of •OH, the E-Peroxone process may also reduce the formation of toxic byproducts. For example, Wang *et al.* found that the bromate formed in the E-Peroxone process was less than one-fourth of that in ozonation, probably due to the decomposition of O_3 and the reduction of hypobromous acid back to bromide by electro generated H_2O_2 (Wang *et al.* 2018). Thus, by simply installing electrodes in ozone contactors, the E-Peroxone process can significantly improve the removal of recalcitrant contaminants. Therefore, it is considered as a promising AOP thanks to the easiness in operation and scale-up and exclusion of chemical dosage as well (Zhan *et al.* 2016).

The electrode material plays a key role in E-Peroxone. Compared with the rare metal, carbon materials have been widely used as electrodes due to their characteristics such as the non-toxic, better chemical stability and corrosion resistance (Mounfield *et al.* 2018; An *et al.* 2019). The frequently used carbon materials as E-Peroxone electrodes are carbon nanotubes (Yang *et al.* 2018; Li *et al.* 2020) and activate carbon fibers (Zhan *et al.* 2016), and carbon/graphite felt (Liu *et al.* 2017; Yang *et al.* 2017a). The efficiency of the carbon materials can be further enhanced with polymeric substances like polytetra-fluoroethylene (PTFE) and dimethyl silicon oil (DMS) (Xu *et al.* 2019). However, the above carbon materials are faced with the problems like high cost and non-green production due to the usage of non-renewable feedstock. Some studies have been conducted to make electrodes with waste-based carbon made from biomass (Huggins *et al.* 2014) and waste paper (Liu *et al.* 2012). However, to our best knowledge, waste-tire carbon (WTC), a carbon material made from the vast volume of waste tires, has not been studied as electrodes in E-Peroxone. The global increment of waste tires is approximately around 17 million tons per year now and would be up to 1.2 billion tons per year in 2030 s (Xu *et al.* 2021a). A potential solution for the waste tires is to produce carbon materials with the pyrolysis process. In addition, WTC also contains elements like N, O, S (San Miguel *et al.* 2003; Xu *et al.* 2021b), which are beneficial for the electrogeneration of H_2O_2 (Yang *et al.* 2017b; Ding *et al.* 2021).

Conversely, antibiotics are widely applied in clinics and animal husbandry to treat bacterial infections. Tinidazole (TNZ), a typical nitroimidazole antibiotic, is often used to prevent or treat amoeba infections, vaginal trichomoniasis and giardiasis (Qin *et al.* 2020). After administration, it could be excreted and enter the ecosystem via sewage, surface runoff, or soil infiltration. Consequently, TNZ has been detected in many water bodies including tap water, e.g. 1.8–17.8 ng/L in a report (Zhao *et al.* 2016). Researchers also found that TNZ may cause genotoxicity and cytotoxicity (Lopez Nigro & Carballo 2008). Therefore, it is vital to choose some efficient water-treatment technologies for the reduction of TNZ in the aquatic environment. TNZ is not highly reactive with O_3 (kinetic constant $330\,M^{-1}\,s^{-1}$) (Rivera-Utrilla *et al.* 2010). Therefore E-Peroxone could be an effective alternative.

The objective of this study was to investigate, for the first time, the feasibility of modifying the graphite felt (GF) with WTC to enhance the performance of E-Peroxone in the degradation of TNZ as a target contaminant. PTFE was also applied to develop a composite electrode (WTC/PTFE-GF) whose generation efficiency of H_2O_2 was specially studied. Some operational parameters were investigated to identify the optimal condition. We also studied the degradation mechanism of TNZ, including the contribution of reactive oxygen species (ROS), the transformation products (TPs), and the degradation pathways.

2. MATERIALS AND METHODS

2.1. Chemicals

WTC was provided by Rixin Hengli Rubber & Plastic Co. Ltd, and carbon black (Li-2060) was purchased from Cyber Electrochemical Materials Corporation. Tinidazole, PTFE (60 %wt.) and sodium indigotin-disulfonate were purchased from Aladdin; sodium chloride, sodium sulfate, anhydrous ethanol, potassium iodide, and tert-butanol (TBA) from Sinopharm Chemical Reagent Co., Ltd; sodium hydroxide, nitric acid, sodium thiosulfate, chloroform ($CHCl_3$), titanium potassium oxalate and sulfuric acid from Shanghai Lingfeng Chemical Reagent Co., Ltd. Except formic acid and acetonitrile being high-performance liquid chromatography (HPLC) grade, all other chemicals were at analytic grade. The deionized water was obtained from a pure water-treatment system (EPED-Z1-30T) and used during all the experimental processes.

2.2. Cathode fabrication

The acidification of WTC was conducted by immersing 5 g WTC in 250 mL solution of 12 M HNO_3 for one hour. The round shapes of GF sheets with a diameter of 50 mm were firstly pretreated with a mixture of 30 mL acetone and 30 mL deionized water for 30 min. Then three pretreated GF sheets were separately immersed into the precursor mixture of 0.3 mL 60% PTFE emulsion, 60 mL ethanol and 1.2 g acidified WTC powder with carbon black powder (mass ration being 1:1), or 0.6 g acidified WTC powder with carbon black powder (mass ration being 1:1), or 0.6 g carbon black powder, named as 1.2 gWTC/PTFE-GF, 0.6 gWTC/PTFE-GF and 0.6 gCB/PTFE-GF, respectively. The three loaded GF sheets were heated in a muffle oven at 350 °C for 60 min. The loaded process was repeated several times until the loading was finally around 10 mg cm^{-2} and no obvious bare surface of the cathode was observed. The anode was made of ruthenium–iridium titanium mesh supplied from Qinghe Yunxun Metallic Materials Co Ltd.

2.3. Experimental

As shown in Figure 1, E-Peroxone reactions were carried out in identical cylinder reactors with an effective volume of 400 mL. Two electrodes (Anode: Ti mesh coated with Ru oxides; Cathode: WTC/PTFE-GFs) in the same shape (50 mm in diameter) were fixed in parallel at a distance of 15 mm in the reactor and connected onto a DC power source (ITECH, IT6721). Next, 50 mM Na_2SO_4 was used as an electrolyte except the test for chlorine effect in which Na_2SO_4 was partly replaced with NaCl to maintain the same conductivity.

The different gaseous ozone dosages were generated with a constant flow rate of 1 L min^{-1} from a laboratory-scale ozone generator (Anseros, COM-AD) and was introduced via a diffusor at the reactor bottom. The degradation efficiency of TNZ was monitored along the reaction under various conditions, such as different ozone concentration, current density, initial pH values and electrolyte. The exhaust gas including O_2/O_3 was absorbed with a saturated solution of KI. To study the effects of ROS, varying dosages of TBA and $CHCl_3$ were applied as scavengers for $\cdot OH$ and $O_2^{\cdot-}$, respectively.

2.4. Analytical method

The applied ozone concentrations were determined in aqueous phase by indigo disulfonate spectrophotometry at 610 nm. The concentration of H_2O_2 was measured using a potassium titanium oxalate method with a spectrophotometer (UV-1900PC, AOE Instruments, China) at 400 nm. The chemical oxygen demand (COD) was measured with a quick photometric method with potassium permanganate (China Standard GB 11892-89).

The current efficiency of H_2O_2 generation was calculated using the following equation (Equation (3)):

$$\eta_i = n \cdot F \cdot [H_2O_2] \cdot V/I \cdot t \cdot 100\% \tag{3}$$

where n is the number of electrons transferred for oxygen reduction to H_2O_2 ($n = 2$), F the Faraday constant (96,486 C/mol), $[H_2O_2]$ the concentration of H_2O_2 (mol/L), V the volume of the solution (L), I the applied current (A), and t is the electrolysis time (s).

TNZ concentration was determined by HPLC (Agilent 1260) equipped with a UV-visible detector and an Infinitylab Poroshell 120 EC-C18 column (2.7 μm particle size; 4.6 × 100 mm). The mobile phase consisted of 70% formic acid (0.1%, v/v) and 30% acetonitrile at a flow rate of 0.35 mL min^{-1}. Injection volume was 20 μL and the detector wavelength was set at 318 nm. The apparent kinetic constant of pseudo first order of TNZ removal (K) was calculated using the following formula

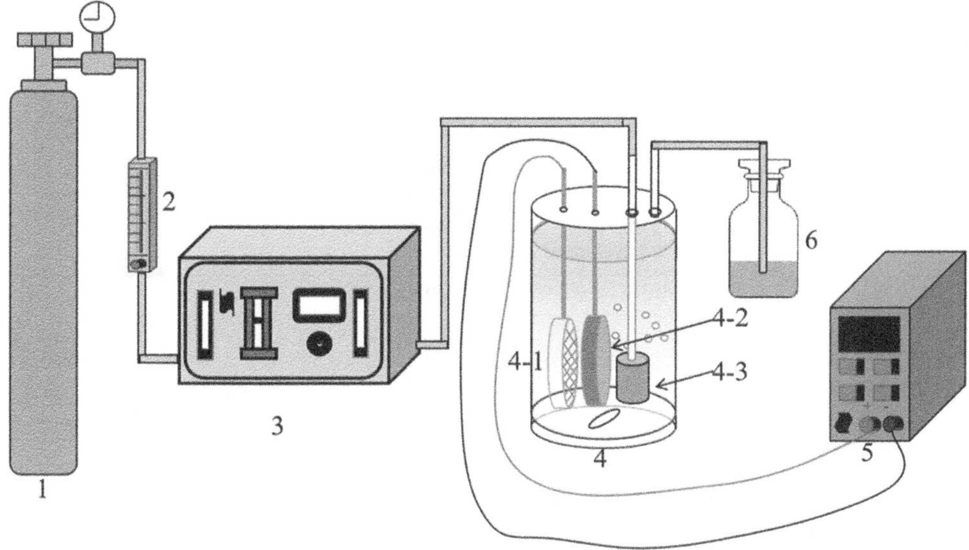

Figure 1 | The set-up diagram of the E-Peroxone system (1-Oxygen tank; 2-Gas flow meter; 3-Ozone generator; 4-Catalytic reactor; 4-1 Anode: RuO_2/Ti; 4-2 Cathode; 4-3 Aerator; 5-DC power; 6-KI solution).

(Equation (4)):

$$K = -\ln (C/C_0)/t \tag{4}$$

where C_0 and C denote the concentration of the TNZ solution at the beginning and at the given time t, respectively.

The TPs of TNZ were analyzed by UPLC-MS (Dinonex Ultimate 3000 UPLC) equipped with C18 column (100 mm × 2.1 mm, 1.9 µm) (Hypersil GOLD, Thermo Scientific, USA). The mobile phase was operated in a gradient mode at a flow of 0.2 mL min^{-1} (20% B at 0 min., 100% B at 8 min., 20% B at 8.1 min.) where Channel A was water with 0.1% formic acid; and Channel B was acetonitrile with 0.1% formic acid. The sample injection volume was 5 µL. The UPLC system was coupled with a Thermo Scientific Q Exactive high-resolution mass spectrometer with electrospray ionization (ESI). Analytical parameters were determined in a positive ionization mode, acquiring spectra in a mass range between 50 and 750 m/z.

3. RESULTS AND DISCUSSION

3.1. Electrogeneration of H_2O_2 with different cathodes

We analyzed the H_2O_2 concentration in 30 min with different cathodes, including 0.6 gCB/PTFE-GF, 0.6 gWTC/PTFE-GF and 1.2 gWTC/PTFE-GF, in 50 mM Na_2SO_4 solutions at pH 7.0, a value of most wastewater and water bodies. The current density was set as 10 mA cm^{-2}. Figure 2(a) illustrates the profiles of H_2O_2 concentrations. It can be seen that H_2O_2 concentrations in all tests were gradually increased up along the reaction time. The electrodes doped with WTC produced more H_2O_2 than that with CB at the same loading: 65 mg L^{-1} vs. 23 mg L^{-1} in 30 min. When the loading of WTC was doubled from 0.6 g to 1.2 g, the production of H_2O_2 was increased from 65 mg L^{-1} and 86 mg L^{-1} in 30 min. In addition, the corresponding current efficiencies of the 0.6 gCB/PTFE-GF, 0.6 gWTC/PTFE-GF and 1.2 gWTC/PTFE-GF cathode were 15.1%, 40.6% and 54.7%. Therefore, with the same dosage in the cathodes, the yield of H_2O_2 and the current efficiency with WTC were around 2.7 times of the data with CB. In Figure 2(b), we can find that the current efficiency was gradually dropping down along the reaction. This might be a result of the breakdown of excessive H_2O_2 at the cathode via a four-electron reaction (Yang *et al.* 2021).

The superior performance of WTC could be possibly attributed to the elements such as N and S in WTC which is absent in CB (San Miguel *et al.* 2003; Xu *et al.* 2021b). As shown in Figure 3, the modified WTC contained 0.39% S and 1.10% N. Nevertheless, further investigation are required to clarify the precise contribution of those elements.

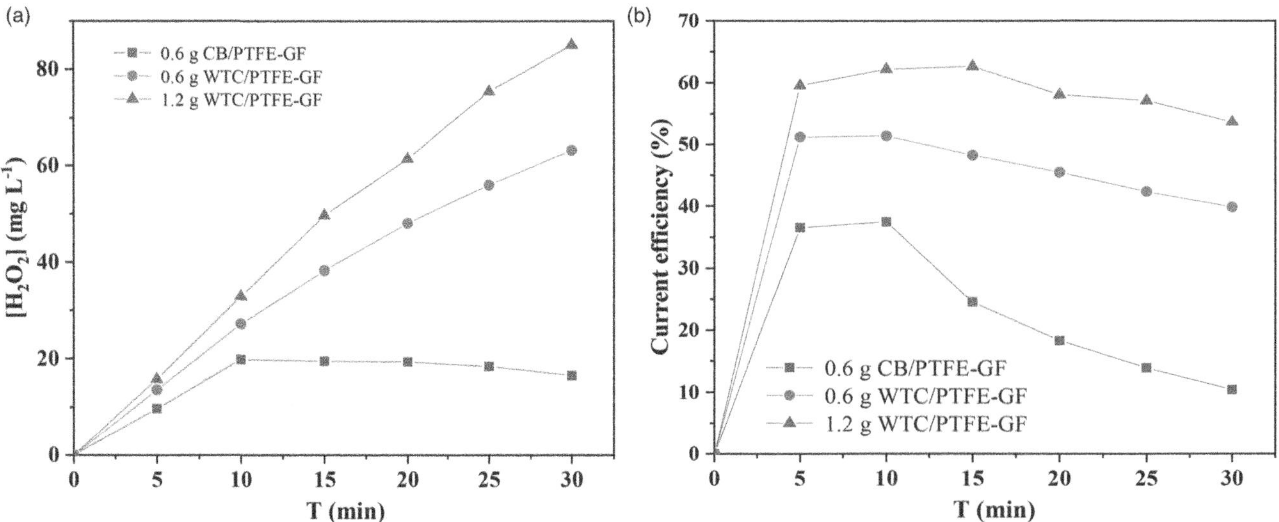

Figure 2 | Evolution of H_2O_2 concentration (a) and its current efficiency (b) during the electrolysis of 400 mL of a 50 mM Na_2SO_4 solution at pH 7.0 with the current density of 10 mA cm^{-2} at room temperature with the different modified cathodes.

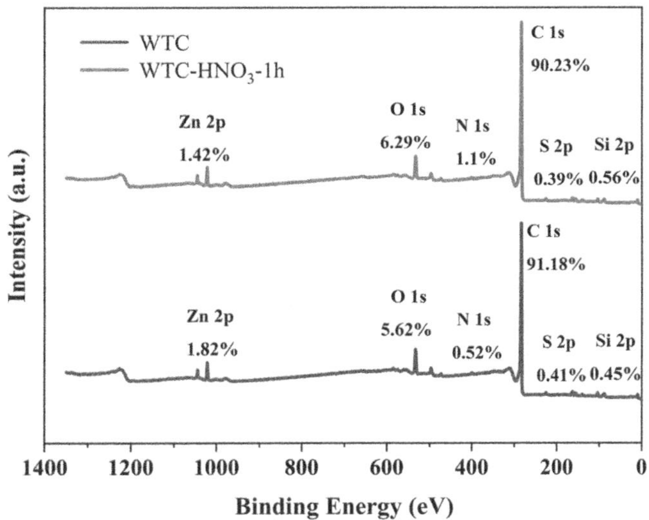

Figure 3 | XPS spectra of pristine and modified WTC.

3.2. E-Peroxone and ozonation

Since H_2O_2 can be generated on the cathode in E-Peroxone and then ROS can be initiated, a better degradation of TNZ can be expected, as demonstrated in previous studies (Mao *et al.* 2018). As shown in Figure 4, when 20 mA cm^{-2} current was applied, TNZ removal was dramatically enhanced at all studied ozone concentrations. For instance, at 11.3 mg L^{-1} gaseous ozone, the removal efficiencies of TNZ in 15 min were increased from 99.5% to 99.90% and the corresponding first-order-kinetic constant increased from 0.35 min^{-1} to 0.41 min^{-1} when the current was switched on. As an essential reactant, ozone can effectively regulate the removal of TNZ in both processes. For example, in the E-Peroxone process, the increase of gaseous ozone from 7.8 mg L^{-1} to 20.6 mg L^{-1} elevated the removal of TNZ from 73% to 100% in 15 min, and the corresponding kinetic constant from 0.085 min^{-1} to 0.78 min^{-1}. Nevertheless, it can be found that the enhancement of TNZ removal with electrochemistry was more significant at the low level of O_3 (e.g. 7.8 mg L^{-1}) than the high level of O_3 (e.g. 20.6 mg L^{-1}). This was probably because the removal of TNZ came from two reactions: O_3 and ·OH formed from

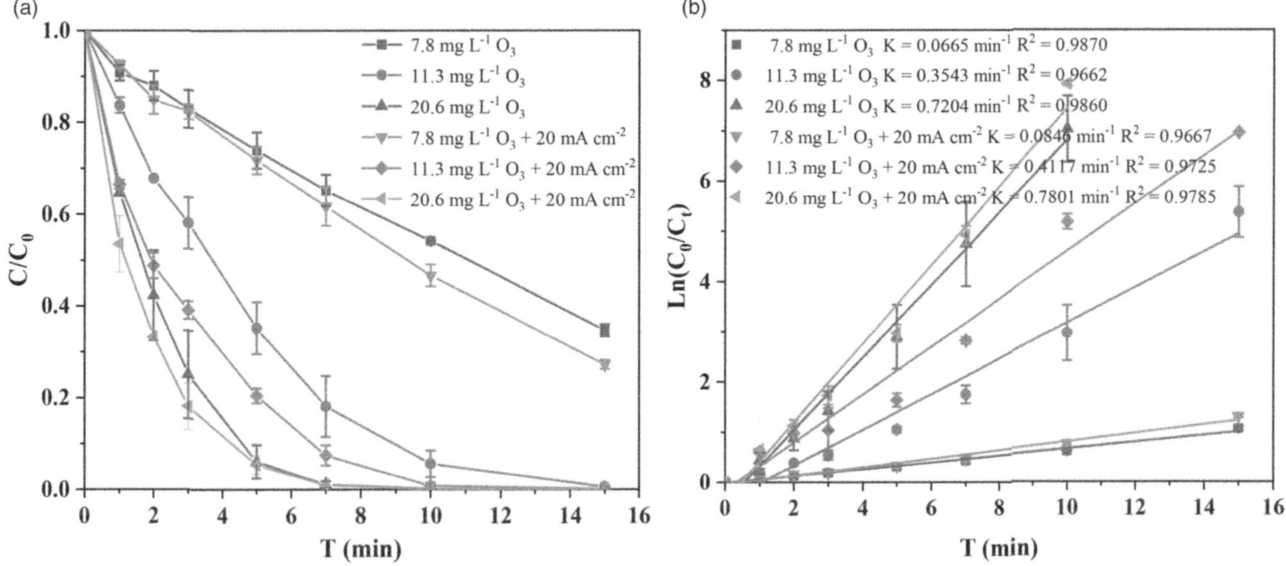

Figure 4 | Performance comparison of E-Peroxone with the individual ozonation process on the TNZ removal in different ozone concentrations. (Conditions: flow rate of O_2/O_3 1 L min^{-1}, 50 mM Na$_2$SO$_4$, initial TNZ: 100 mg L^{-1}, pH = 7.0. Error bars represent the standard error.)

O_3/H_2O_2. When O_3 was increased, its contribution to the removal of TNZ can also be enhanced while the yield of H_2O_2 was likely to be constant at the same current density. Furthermore, the excessive O_3 may react with H_2O_2 and an ineffective decomposition of ozone may occur (Cornejo & Nava 2021). Therefore, it is important to match the yield of H_2O_2 and ozone dosage.

Current density is another important factor regulating the generation of H_2O_2 and subsequently may strongly influence the contaminant degradation in E-Peroxone. Figure 5 shows the effect of current density varying from 0 to 30 mA cm^{-2} on the degradation of TNZ in E-Peroxone. When the high current density was applied, the degradation of TNZ increased correspondingly. This trend can be attributed to the fact that the high applied current density can enhance the electrogeneration of H_2O_2, which commonly happens in other studies (Valim *et al.* 2021). Consequently, more •OH could be produced to degrade TNZ at higher applied currents. As the generation of H_2O_2 is the key parameter in E-Peroxone, the low current density, 10 mA cm^{-2} in this case, would generate a low level of H_2O_2 and thus result in an insignificant enhancement of TNZ removal in all dosages of ozone (Wang *et al.* 2018). By increasing the current density to more than 20 mA cm^{-2}, a catalytic effect can be clearly seen, especially at the gaseous ozone concentrations of 7.8 mg L^{-1} and 11.3 mg L^{-1}. Nevertheless, when the gaseous ozone concentration was increased to 20.6 mg L^{-1}, the catalytic effect became negligible. The reasons should be similar to the discussion above on the results shown in Figure 4.

The energy consumption can be roughly estimated. Based on the scenario of 7.8 mg L^{-1} and 1 L min^{-1} ozone (468 mg h^{-1}), the energy consumption for ozone generation from air would be around 23 Wh (20 g kWh^{-1} at the industrial scale). The generation of H_2O_2 (5 V, 30 mA cm^{-2}) would consume 3 Wh, accounting for approximately 13% of the energy of ozone production. Nevertheless, when oxygen is used as a feed gas, the energy efficiency of ozonation generation could be further reduced.

3.3. Effect of initial pH

The solution pH plays an important role in the peroxone process which is usually initiated with the reaction of O_3 and HO_2^-, the anion of H_2O_2, with the generation of $HO_2^{\bullet}$ and $O_3^{\bullet -}$ (Equations (5) and (6)) (Von Sonntag & Von Gunten 2012). $HO_2^{\bullet}$ can further decompose to $O_2^{\bullet -}$ while $O_3^{\bullet -}$ could be protonated to $HO_3^{\bullet}$ which can lead to •OH (Equations (7)–(9)):

$$H_2O_2 \rightleftharpoons HO_2^- + H^+ \quad (pK_a = 11.8) \tag{5}$$

$$O_3 + HO_2^- \rightarrow HO_2^{\bullet} + O_3^{\bullet -} \tag{6}$$

$$HO_2^{\bullet} \rightleftharpoons O_2^{\bullet -} + H^+ \quad (pK_a = 4.8) \tag{7}$$

$$O_3^{\bullet -} + H^+ \rightleftharpoons HO_3^{\bullet} \tag{8}$$

$$HO_3^{\bullet} \rightleftharpoons {}^{\bullet}OH + O_2 \quad (pK_a = 8.2) \tag{9}$$

The effects of initial pH values (pH 3-11) were investigated on the degradation of TNZ. To avoid the interference of inorganic ions, no buffer was involved in the tests. As shown in Figure 6(a), a higher removal of TNZ was obtained at higher pH values at the beginning stage (<3 min). Finally, a comparable removal efficiency was obtained after 5 min in all pH values. This may be attributed to the similar final pH values which were around 3.5 (Figure 6(b)). It is widely recognized that some acidic molecules can be formed during the oxidation of organic contaminants in the ozone-based AOPs leading the declination of pH to the range of 2.9–4.1 (Cornejo *et al.* 2021). As shown in Equations (5)–(9), the low pH value may impair the formation of ROS.

3.4. Effect of initial TNZ concentration

The trends of TNZ degradation with a function of its initial concentrations from 40 to 120 mg L^{-1} were investigated with 50 mM Na_2SO_4, pH 7.0, the current density of 20 mA cm^{-2}, and gaseous O_3 dose of 10 mg L^{-1}. Figure 7 shows that the TNZ degradation occurred promptly and nearly complete removal of TNZ was achieved in 5–7 min. The corresponding apparent kinetic constants decreased slightly with the increase in initial TNZ concentration. The best kinetic constant of the TNZ degradation was obtained as the initial TNZ concentration of 40 mg L^{-1}: 1.37 min^{-1} (R^2 = 0.96). The phenomenon might be related to the limited dissolved O_3. The high initial concentration of TNZ can consume more O_3 and the dissolved O_3 level would become limited when the mass transfer of ozone is constant. Moreover, the degradation of TNZ would generate TPs which could further complete for O_3 and ˙OH and then lead to a low degradation kinetic. For instance, after

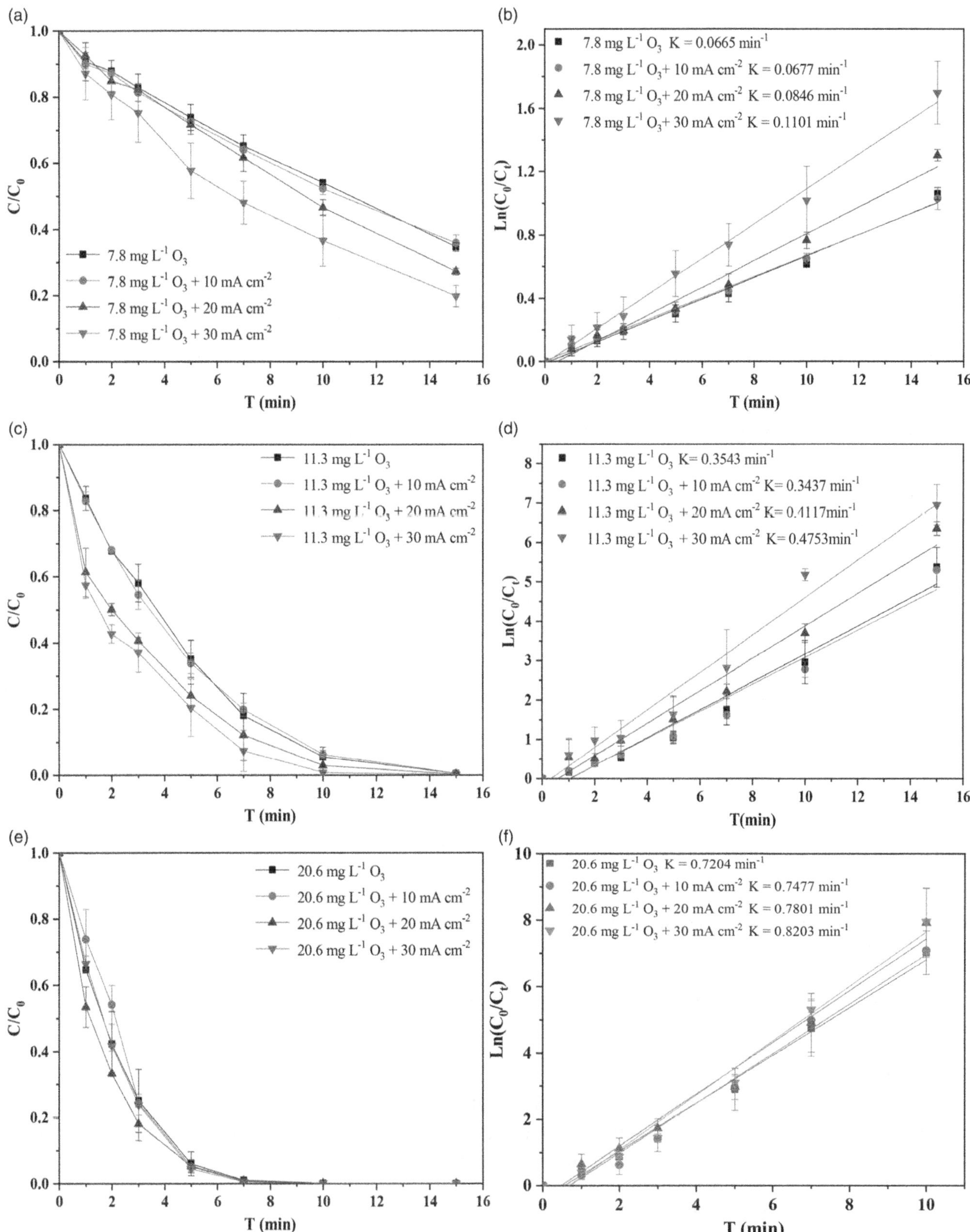

Figure 5 | TNZ removal of E-Peroxone and individual ozonation process as a function of applied current density at different gaseous ozone levels. (Conditions: low rate of O_2/O_3 1 L min^{-1}, 50 mM Na_2SO_4, 100 mg L^{-1} TNZ, pH = 7. Error bars represent the standard error.)

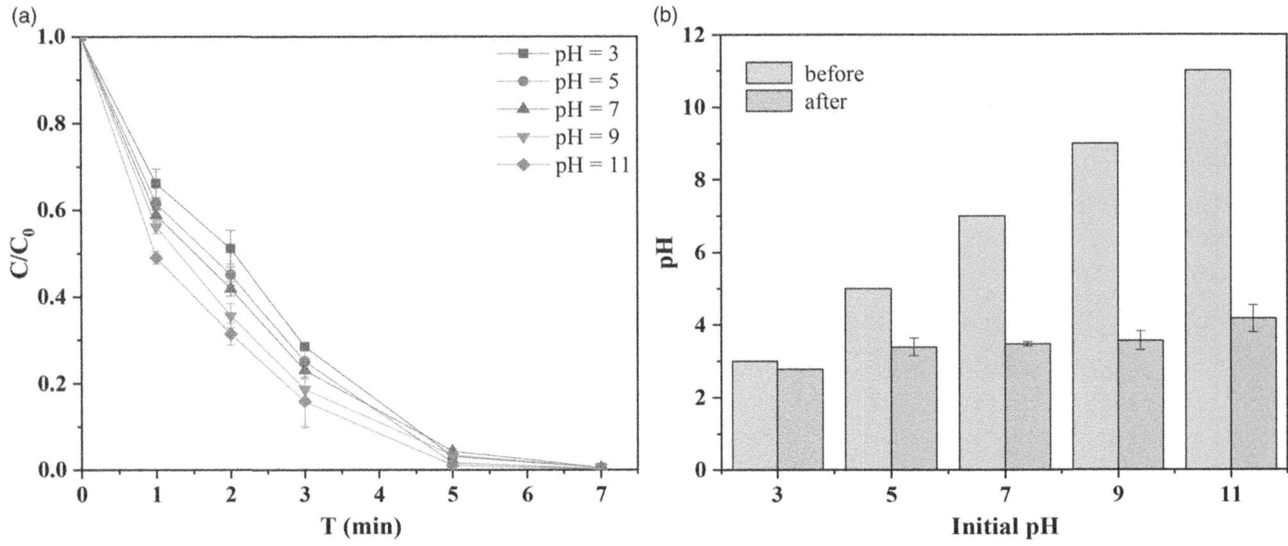

Figure 6 | Effect of pH values on TNZ degradation (a) and pH change before and after the E-Peroxone process (b). (Conditions: 20.6 mg L^{-1} O$_3$ and current density 20 mA cm^{-2}, 50 mM Na$_2$SO$_4$, 100 mg L^{-1} TNZ).

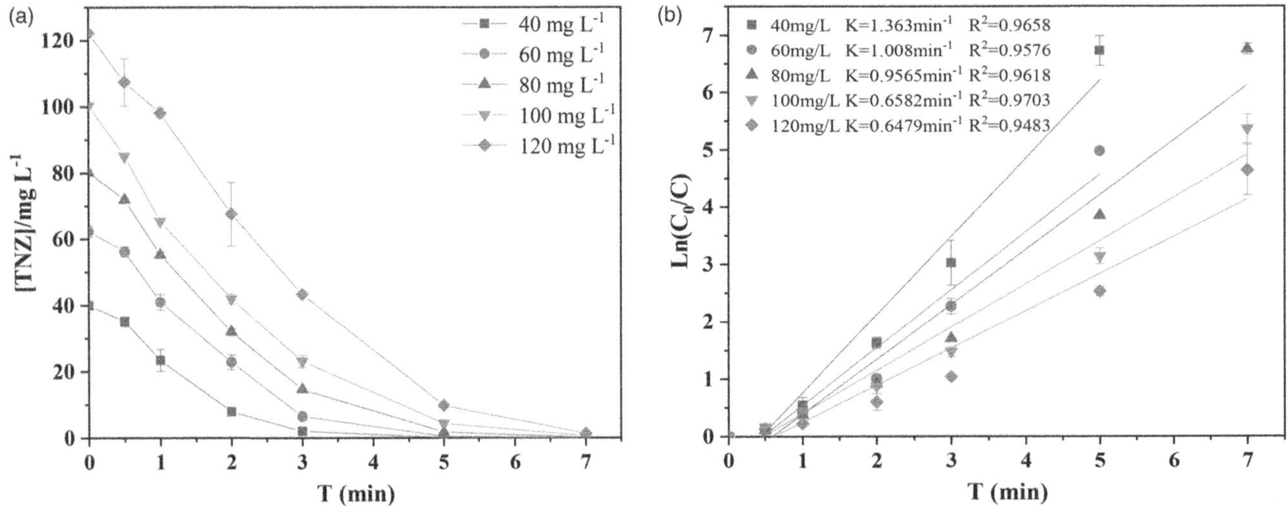

Figure 7 | The degradation of TNZ at different initial concentrations (Conditions: 20.6 mg L^{-1} O$_3$ and current density 20 mA cm^{-2}, 50 mM Na$_2$SO$_4$, pH = 7.0).

5 min, there was still 89 mg/L COD in the test of 120 mg/L TNZ, several folds higher than the COD (24 mg/L) in the test of 40 mg/L TNZ.

3.5. Effect of chloride

Wastewater usually contains various components that may influence the degradation of contaminants in E-Peroxone. We varied the content of chloride ions from 0 to 50 mM and sulfate ion ranged from 0 to 70 mM and finally the same conductivity was kept in solution. A significant inhibition of chloride on TNZ degradation, however, was not observed, as shown in Figure 8. This means that E-Peroxone may be suitable for the treatment of chlorine-containing wastewater. Chloride ions play a complex role in E-Peroxone. They can be oxidized at the anode with the formation oxychlorine anions such as hypochlorite (ClO$^-$), chlorite (ClO$_2^-$), chlorate (ClO$_3^-$) and perchlorate (ClO$_4^-$), which may influence E-Peroxone by consuming H$_2$O$_2$ and O$_3$ (Equations (10)–(14)) (Lin *et al.* 2016). Conversely, The Cl$^{\cdot}$/ClO$_2^{\cdot}$ radicals can also be produced at the anode via Equations (15) and (16) (Ganiyu & Gamal El-Din 2020). Both radicals may contribute to the oxidation of organic

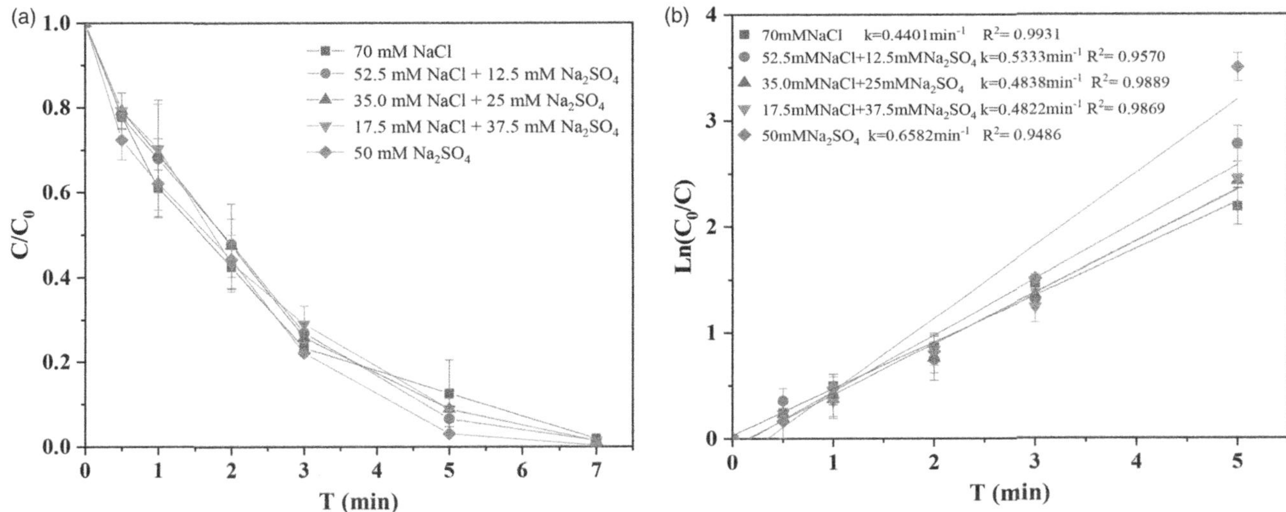

Figure 8 | The effect of electrolytes on the degradation of TNZ. (Conditions: 20.6 mg L^{-1} O_3 and current density 20 mA cm^{-2}, pH = 7.0, 100 mg L^{-1} TNZ.)

contaminants and thus may compensate for the negative impact of consuming oxidants by chlorine (Lei *et al.* 2021):

$$2ClO^- + 2H_2O_2 \rightarrow 2Cl^- + 3O_2 + 4H^+ \tag{10}$$

$$2H^+ + 2ClO_3^- + H_2O_2 \rightarrow 2ClO_2 + 2H_2O + O_2 \tag{11}$$

$$ClO^- + O_3 \rightarrow O_2 + ClO_2^- \tag{12}$$

$$ClO_2^- + O_3 \rightarrow O_3^- + ClO_2 \tag{13}$$

$$ClO_2^- + O_3 \rightarrow O_2 + ClO_3^- \tag{14}$$

$$Cl^- \rightarrow Cl^\bullet + e^- \tag{15}$$

$$ClO_2^- \rightarrow ClO_2^\bullet + e^- \tag{16}$$

3.6. Roles of ROS

We applied TBA and $CHCl_3$ as scavengers to investigate the effects of $\bullet OH$ and $O_2^{\bullet-}$. TBA is unreactive to ozone ($k = 2 \times 10^{-5}$ M^{-1} s^{-1}) and $O_2^{\bullet-}$ but can be easily oxidized by $\bullet OH$ ($k = 6 \times 10^8$ M^{-1} s^{-1}). Therefore it is often used to quench the reaction of $\bullet OH$ without affecting the reaction of O_3 (Guo *et al.* 2022). $CHCl_3$ is also recalcitrant to ozone ($k < 0.05$ M^{-1} s^{-1}) and reacts more quickly with $O_2^{\bullet-}$ ($k = 2.3 \times 10^8$ M^{-1} s^{-1}) than $\bullet OH$ ($k = 5.4 \times 10^7$ M^{-1} s^{-1}). Different amounts of TBA and $CHCl_3$ were added into the reaction solution. It was found that the addition of TBA clearly inhibited the degradation: the pseudo-first-order kinetic constant was nearly halved (51%) with 20 mM TBA (Figure 9(a)), suggesting the participation of $\bullet OH$. A higher inhibition effects occurred with the presence of the same levels of $CHCl_3$ (Figure 9(b)). Here, 20 mM $CHCl_3$ depressed the pseudo-first-order kinetic constant by 61%, compared with the control test. The higher inhibition effect of $CHCl_3$ might be related to the simultaneous quenching of $\bullet OH$ and $O_2^{\bullet-}$. Moreover, quenching $O_2^{\bullet-}$ by $CHCl_3$ can also result in a low generation of $O_3^{\bullet-}$ (Equation (17)) which is an important precursor for $\bullet OH$ as shown in Equations (8) and (9).

$$O_2^{\bullet-} + O_3 \rightarrow O_2 + O_3^{\bullet-} \tag{17}$$

3.7. Degradation pathway of TNZ

In the oxidation processes, the organic contaminants are usually oxidized into some TPs before the complete mineralization. The TPs of TNZ in E-Peroxone were analyzed with the UPLC-MS system. According to the m/z data and the literature information (Acosta-Rangel *et al.* 2018), the molecular structures of five TPs can be estimated (Table 1), based on which of the

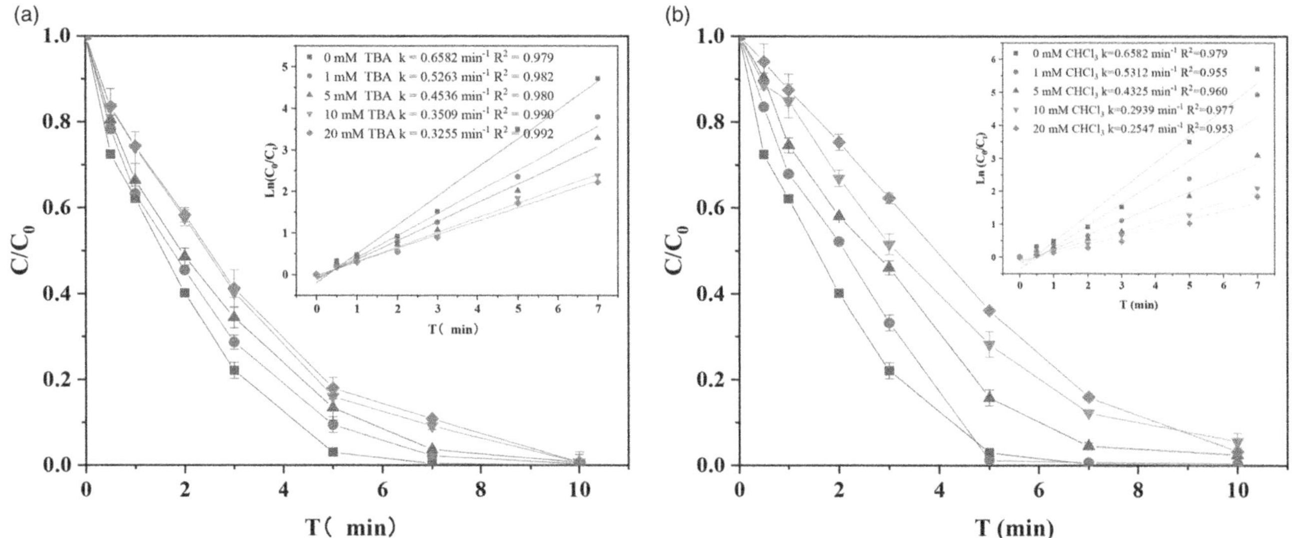

Figure 9 | TNZ degradation with different scavenger concentrations of TBA (a) and CHCl$_3$ (b). (Conditions: 20.6 mg L^{-1} O$_3$ and current density of 20 mA cm^{-2}, 100 mg L^{-1} TNZ, pH = 7.0, 50 mM Na$_2$SO.)

Table 1 | Intermediate products of TNZ during E-Peroxone process

	Mass (m/z)	Elemental formula	Proposed structure
TNZ	247.06	C$_8$H$_{13}$N$_3$O$_4$S	
TP1	263.06	C$_8$H$_{13}$N$_3$O$_5$S	
TP2	236.28	C$_8$H$_{16}$N$_2$O$_4$S	
TP3	281.06	C$_8$H$_{15}$N$_3$O$_6$S	
TP4	178.08	C$_6$H$_{14}$N$_2$O$_2$S	
TP5	164.06	C$_5$H$_{12}$N$_2$O$_2$S	

degradation pathways were proposed. As shown in Figure 10, the degradation may possibly start with the hydroxylation on the N-heterocyclic ring with the formation of TP1. Further hydroxylation may continue to form TP2. The nitro group of TP1 can also likely be replaced with a hydroxyl group to generate TP3. Subsequently, the opening of the N-heterocyclic ring and the loss of methyl group could finally lead to TP4 and TP5.

4. CONCLUSIONS

In this study, we innovatively applied WTC as a recycled material to fabricate a cathode for the E-Peroxone process and compared its performance with commercial carbon black. The results demonstrated that the addition of WTC can significantly

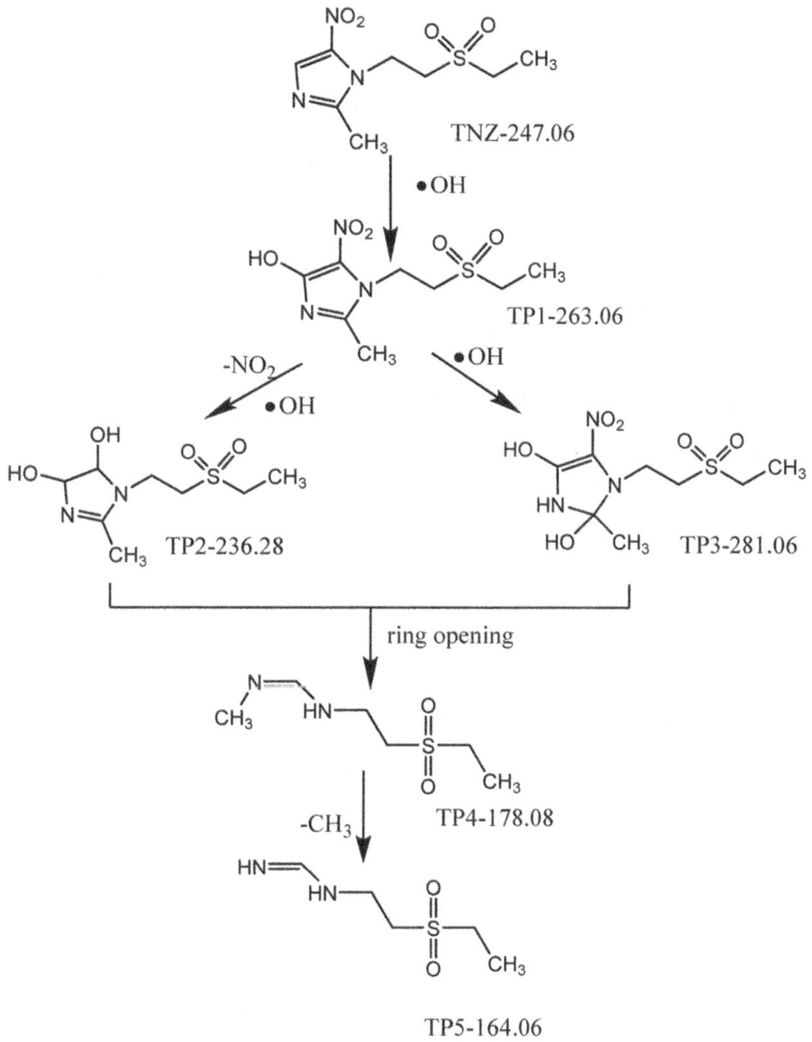

Figure 10 | Proposed pathways of TNZ degradation.

enhance the current efficiency of generating H_2O_2, up to 54.7%, which was about 2.7 times that of commercial carbon black at the same loading. For the efficient degradation of TNZ, the dosage of ozone should be fine tuned with the current density. In the study, the significant interference of chlorine (up to 50 mM) was not noticed. The main ROS were probably •OH and $O_2^{\bullet-}$, as indicated by the scavenger tests. The degradation of TNZ may involve the hydroxylation and the opening of the N-heterocyclic ring, as the identified TPs indicated. In general, WTC can be a potential and economic material to replace commercial carbon for the construction of cathodes used in E-Peroxone. However, further investigations should be conducted to evaluate the long-term stability of the WTC cathode and to estimate the construction and operation cost.

ACKNOWLEDGEMENTS

The study was funded by the National Natural Science Foundation of China (Grant No. 52070097) and the Natural Science Foundation of Jiangsu Province, China (Grant No. BK20200699).

DATA AVAILABILITY STATEMENT

All relevant data are included in the paper or its Supplementary Information.

CONFLICT OF INTEREST STATEMENT

The authors declare there is no conflict.

REFERENCES

Acosta-Rangel, A., Sánchez-Polo, M., Polo, A. M. S., Rivera-Utrilla, J. & Berber-Mendoza, M. S. 2018 Tinidazole degradation assisted by solar radiation and iron-doped silica xerogels. *Chemical Engineering Journal* **344**, 21–33.

An, J., Li, N., Zhao, Q., Qiao, Y., Wang, S., Liao, C., Zhou, L., Li, T., Wang, X. & Feng, Y. 2019 Highly efficient electro-generation of H_2O_2 by adjusting liquid-gas-solid three phase interfaces of porous carbonaceous cathode during oxygen reduction reaction. *Water Research* **164**, 114933.

Brillas, E. 2022 A critical review on ibuprofen removal from synthetic waters, natural waters, and real wastewaters by advanced oxidation processes. *Chemosphere* **286** (Pt 3), 131849.

Cornejo, O. M. & Nava, J. L. 2021 Mineralization of the antibiotic levofloxacin by the electro-peroxone process using a filter-press flow cell with a 3D air-diffusion electrode. *Separation and Purification Technology* **254**, 117611.

Cornejo, O. M., Ortiz, M., Aguilar, Z. G. & Nava, J. L. 2021 Degradation of Acid Violet 19 textile dye by electro-peroxone in a laboratory flow plant. *Chemosphere* **271**, 129804.

Ding, Y., Zhou, W., Gao, J., Sun, F. & Zhao, G. 2021 H_2O_2 electrogeneration from O_2 electroreduction by N-doped carbon materials: a mini-review on preparation methods, selectivity of N sites, and prospects. *Advanced Materials Interfaces* **8** (10), 2002091.

Ganiyu, S. O. & Gamal El-Din, M. 2020 Insight into in-situ radical and non-radical oxidative degradation of organic compounds in complex real matrix during electrooxidation with boron doped diamond electrode: a case study of oil sands process water treatment. *Applied Catalysis B: Environmental* **279**, 119366.

Guo, Y., Long, J., Huang, J., Yu, G. & Wang, Y. 2022 Can the commonly used quenching method really evaluate the role of reactive oxygen species in pollutant abatement during catalytic ozonation? *Water Research* **215**, 118275.

Huggins, T., Wang, H., Kearns, J., Jenkins, P. & Ren, Z. J. 2014 Biochar as a sustainable electrode material for electricity production in microbial fuel cells. *Bioresource Technology* **157**, 114–119.

Lei, Y., Lei, X., Westerhoff, P., Zhang, X. & Yang, X. 2021 Reactivity of chlorine radicals (Cl(*) and Cl_2(*-)) with dissolved organic matter and the formation of chlorinated byproducts. *Environmental Science & Technology* **55** (1), 689–699.

Li, Z., Shen, C., Liu, Y., Ma, C., Li, F., Yang, B., Huang, M., Wang, Z., Dong, L. & Wolfgang, S. 2020 Carbon nanotube filter functionalized with iron oxychloride for flow-through electro-Fenton. *Applied Catalysis B: Environmental* **260**, 118204.

Lin, Z., Yao, W., Wang, Y., Yu, G., Deng, S., Huang, J. & Wang, B. 2016 Perchlorate formation during the electro-peroxone treatment of chloride-containing water: effects of operational parameters and control strategies. *Water Research* **88**, 691–702.

Liu, M.-C., Kong, L.-B., Lu, C., Li, X.-M., Luo, Y.-C. & Kang, L. 2012 Waste paper based activated carbon monolith as electrode materials for high performance electric double-layer capacitors. *RSC Advances* **2** (5), 1890–1896.

Liu, S., Wang, Y., Wang, B., Huang, J., Deng, S. & Yu, G. 2017 Regeneration of Rhodamine B saturated activated carbon by an electro-peroxone process. *Journal of Cleaner Production* **168**, 584–594.

Lopez Nigro, M. M. & Carballo, M. A. 2008 Genotoxicity and cell death induced by tinidazole (TNZ). *Toxicology Letters* **180** (1), 46–52.

Mao, Y., Guo, D., Yao, W., Wang, X., Yang, H., Xie, Y. F., Komarneni, S., Yu, G. & Wang, Y. 2018 Effects of conventional ozonation and electro-peroxone pretreatment of surface water on disinfection by-product formation during subsequent chlorination. *Water Research* **130**, 322–332.

Mounfield, W. P., Garg, A., Shao-Horn, Y. & Román-Leshkov, Y. 2018 Electrochemical oxygen reduction for the production of hydrogen peroxide. *Chem* **4** (1), 18–19.

Qin, Q., Qin, H., Li, K., Tan, R., Liu, X. & Li, L. 2020 The adsorption characteristics and degradation mechanism of tinidazole on an anatase TiO_2 surface: a DFT study. *RSC Advances* **10** (4), 2104–2112.

Rivera-Utrilla, J., Sanchez-Polo, M., Prados-Joya, G., Ferro-Garcia, M. A. & Bautista-Toledo, I. 2010 Removal of tinidazole from waters by using ozone and activated carbon in dynamic regime. *Journal of Hazardous Materials* **174** (1–3), 880–886.

San Miguel, G., Fowler, G. D. & Sollars, C. J. 2003 A study of the characteristics of activated carbons produced by steam and carbon dioxide activation of waste tyre rubber. *Carbon* **41** (5), 1009–1016.

Turkay, O., Ersoy, Z. G. & Barışçı, S. 2017 Review – the application of an Electro-Peroxone process in water and wastewater treatment. *Journal of The Electrochemical Society* **164** (6), E94–E102.

Valim, R. B., Trevelin, L. C., Sperandio, D. C., Carneiro, J. F., Santos, M. C., Rodrigues, L. A., Rocha, R. S. & Lanza, M. R. V. 2021 Using carbon black modified with nb_2o_5 and ruo_2 for enhancing selectivity toward H_2O_2 electrogeneration. *Journal of Environmental Chemical Engineering* **9** (6), 106787.

Von Sonntag, C. & Von Gunten, U. 2012 *Chemistry of Ozone in Water and Wastewater Treatment*. IWA Publishing, London, UK.

Wang, Y., Yu, G., Deng, S., Huang, J. & Wang, B. 2018 The electro-peroxone process for the abatement of emerging contaminants: mechanisms, recent advances, and prospects. *Chemosphere* **208**, 640–654.

Xu, A., He, B., Yu, H., Han, W., Li, J., Shen, J., Sun, X. & Wang, L. 2019 A facile solution to mature cathode modified by hydrophobic dimethyl silicon oil (DMS) layer for electro-Fenton processes: water proof and enhanced oxygen transport. *Electrochimica Acta* **308**, 158–166.

Xu, J., Yu, J., He, W., Huang, J., Xu, J. & Li, G. 2021a Recovery of carbon black from waste tire in continuous commercial rotary kiln pyrolysis reactor. *Science of the Total Environment* **772**, 145507.

Xu, J., Yu, J., He, W., Huang, J., Xu, J. & Li, G. 2021b Wet compounding with pyrolytic carbon black from waste tyre for manufacture of new tyre – a mini review. *Waste Management & Research* **39** (12), 1440–1450.

Yang, W., Zhou, M., Cai, J., Liang, L., Ren, G. & Jiang, L. 2017a Ultrahigh yield of hydrogen peroxide on graphite felt cathode modified with electrochemically exfoliated graphene. *Journal of Materials Chemistry A* **5** (17), 8070–8080.

Yang, Y., He, F., Shen, Y., Chen, X., Mei, H., Liu, S. & Zhang, Y. 2017b A biomass derived N/C-catalyst for the electrochemical production of hydrogen peroxide. *Chemical Communications (Cambridge)* **53** (72), 9994–9997.

Yang, H., Zhou, M., Yang, W., Ren, G. & Ma, L. 2018 Rolling-made gas diffusion electrode with carbon nanotube for electro-Fenton degradation of acetylsalicylic acid. *Chemosphere* **206**, 439–446.

Yang, J., Li, J., Ding, R., Liu, C. & Yin, X. 2021 Kinetic effects of temperature on Fe–N–C catalysts for 2e- and 4e-oxygen reduction reactions. *Journal of The Electrochemical Society* **168** (9), 096502.

Zhan, J., Wang, H., Pan, X., Wang, J., Yu, G., Deng, S., Huang, J., Wang, B. & Wang, Y. 2016 Simultaneous regeneration of p-nitrophenol-saturated activated carbon fiber and mineralization of desorbed pollutants by Electro-Peroxone process. *Carbon* **101**, 399–408.

Zhao, J., He, Q., Zhang, X., Yang, J., Yao, B., Zhang, Q. & Yu, X. 2016 Preparation and properties of visible light responsive g-C$_3$N$_4$/BiNbO$_4$ photocatalysts for tinidazole decomposition. *Integrated Ferroelectrics* **176** (1), 37–53.

First received 14 January 2022; accepted in revised form 27 May 2022. Available online 1 June 2022

doi: 10.2166/wst.2022.071

Fenton oxidation as pretreatment for biomass gasification condensate: cost and biomass inhibition evaluation

Qiqi Zhang, Jonas Pluschke and Sven-Uwe Geißen*

Environmental Process Engineering, Technical University of Berlin, Berlin 10623, Germany
*Corresponding author. E-mail: sven.geissen@tu-berlin.de

ABSTRACT

Gasification transforming organic compounds into energy-rich pyrolytic gas, is a climate-friendly treatment option for biological solid wastes. The condensates arising from the pyrolytic gas valorization is owing to high concentrations of small molecular phenols, cyanides, nitrogen-heterocyclics, aromatics and ammonium, posing an environmental and health hazard. In this paper, the watery phase of the biomass gasification condensate from spent mushroom compost (SMC), with a chemical oxygen demand (COD) of 16.4 g/L and total nitrogen of 2.3 g/L, was pretreated by Fenton oxidation. The experiments were conducted at room temperature with an initial pH value of 3, 5 and 8.9, hydrogen peroxide (H_2O_2) dosages between 15 and 100% of the normalized stoichiometric ratio (NSR), and Fe^{2+} dosages corresponding to molar ratio of H_2O_2:Fe^{2+} between 10 and 30. Through respiration inhibition assays, the best operational condition for detoxification was determined at an initial pH 5 with 30% NSR H_2O_2 dosage and molar ratio of H_2O_2:Fe^{2+} at 15:1. The specific operational cost of the Fenton oxidation was calculated at 2.17 €/kg $COD_{elimination}$. In respiration inhibition assay, the oxygen consumption of wastewater after Fenton oxidation was increased by 316% in 3 days. In a 20 days' biogas production test, the biogas production was increased by 81%.

Key words: cost evaluation, detoxification, gasification condensate, pretreatment by Fenton oxidation, respiration inhibition assay

HIGHLIGHTS

- Detoxification of biomass gasification condensate via Fenton oxidation.
- Modification of respiration inhibition assays with utilization of WTW OxiTop®-C.
- Toxicity evaluation of treated wastewater by respiration inhibition assays.
- Operational parameter optimization based on detoxification capabilities and cost-efficiency.
- Toxicity increased by UV-mediated Fenton oxidation.

1. INTRODUCTION

Gasification is an efficient technology to sustainably dispose of solid biomass (Farzad *et al.* 2016). During the gasification process, the feedstock undergoes a succession of thermal reactions, mainly drying, pyrolysis, combustion, and gasification. The high temperatures in the gasifier (1,100–1,500 °C) guarantee a thorough decomposition of the biomass, such as agriculture wastes, wood waste materials and sewage sludge (Liu et al. 2011; Sansaniwal *et al.* 2017; Widjaya *et al.* 2018).

Spent mushroom compost (SMC) is a waste stream from the industrial mushroom production. Considering a worldwide mushroom production of 35 million tons, an estimated 140–175 million tons of SMC is produced every year (Grimm & Wösten 2018; Umor *et al.* 2021). Due to slow decomposition rates, high content of water and ash as well as a potential pollution to underground and surface water by phosphorus and nitrogen, the SMC is not suitable to be treated by composting, anaerobic digestion, incineration or landfilling (Huang *et al.* 2018). However, the upper calorific value of dried SMC with a value of 11.95 MJ/kg, is comparable to dried sewage sludge, making it suitable for gasification. (Williams *et al.* 2001)

The watery phase of the condensate is produced during the drying of the pyrolytic gas, containing 82–93% water and 7–18% organic compounds (Muzyka *et al.* 2015). Independent of the feed substrate, organic acids, aldehydes, benzenes, ethers, furans, hydrocarbons, ketones and phenols are produced during gasification and pyrolysis processes (Zhang *et al.* 2019a). The recovery of valuable organic compounds from the condensate is limited by its complex composition and low individual concentrations. Moreover, these organic compounds show a moderate or high acute toxicity to the environment and a tendency for bioaccumulation (Ji *et al.* 2016; Huang *et al.* 2019).

Due to the inhibitory effects of the condensate to microorganisms, a conventional treatment of activated sludge processes observed a long lag phase and requires an extended hydraulic retention time, that leads to a higher operation cost (Muzyka *et al.* 2015; Mishra *et al.* 2021). Advanced oxidation processes (AOPs) were widely applied to abate the organic pollutants, such as pharmaceutics, phenols, formaldehyde, dye intermediate in wastewater. (Muzyka *et al.* 2015; Cetinkaya *et al.* 2018; Guo *et al.* 2018; Nidheesh *et al.* 2021). The Fenton oxidation relies on the in-situ production of hydroxyl radical from hydrogen peroxide through catalysis by Fe^{2+} at a low pH value. It is a homogeneous catalytic oxidation process. Owing to the wide availability, low costs and low toxicity, Iron(II) is typically used in the form of iron sulfate as catalyst (Nidheesh *et al.* 2021). Due to a narrow working pH range, high chemical consumption and excessive iron sludge production, numerous optimized Fenton-based AOPs such as UV-Fenton oxidation, heterogeneous catalytic Fenton oxidation and electro-Fenton oxidation have been developed in recent years. However, the UV-Fenton oxidation is limited by light irradiation, rendering it unsuitable for wastewaters with high turbidity or deep color (Brillas 2020). Owing to complicated synthesis routes and high costs of heterogenous catalysts, the heterogeneous Fenton oxidation was often limited to laboratory-scale applications until now (Zhang *et al.* 2019b). The chemical consumption is lowered in electro-Fenton oxidation, but their overall costs are often increased due to the costly electrodes and high electricity demand. A relative COD elimination cost of 5.76 \$/kg was achieved in textile wastewater treatment by electro-Fenton oxidation (Kaur *et al.* 2019).

The complete COD degradation by Fenton oxidation observed a high operation cost, the specific operation cost in biomass gasification wastewater treatment reached US\$8.7/kg COD elimination (Muzyka *et al.* 2015). Through coupling of Fenton oxidation with biological degradation, the treatment process is more cost effective. (Tripathi *et al.* 2013; Zhu *et al.* 2018; Mishra *et al.* 2021). However, generated transformation products and intermediates of the Fenton process may increase the toxicity of the feed (Babu *et al.* 2019). The overall biomass inhibition of wastewater is most efficiently assessed by toxicity assays. Identifying singular components, transformation products and intermediates and evaluating their inhibitory effects individually have been inconclusive (Sharma *et al.* 2018).

In this paper, Fenton oxidation was investigated as a pretreatment for the watery phase of SMC gasification condensate. The operation conditions such as pH value, the dosage of H_2O_2 and Fe^{2+} were optimized in single-factor experiments with regards to detoxification efficiency through respiration inhibition assays (OECD/OCDE 2010; Babu *et al.* 2019). The corresponding removal rates of organic phenolic compounds, COD, and ammonium after Fenton oxidation were also investigated. Moreover, biogas production tests were conducted as an additional evaluation tool for the detoxification effect of Fenton oxidation on the wastewater.

2. METHODS AND MATERIALS

2.1. Wastewater

The counterflow biomass gasification plant located in Westphalia, Germany is fed with SMC (periodically mixed with manure and forestry residues). The pyrolytic gas produced in the gasification process was dried before its valorization in a combined heat and power unit. In the drying process a condensate was segregated by a cyclone water trap. The condensate was separated into a sinking layer with tars and heavy oils, a floating light oil layer, and the watery phase (80–90% water content). The sinking and floating layers were reintroduced into the gasifier, as both mixtures have high calorific value. The remaining condensate of water-soluble components needs to be disposed of before discharge. The overview of the wastewater's main composition is given in Table 1. The wastewater was stored at 6 °C, before experiments the wastewater was pretreated by centrifugation at 4,000 rpm for 20 min followed by membrane filtration (0.45 μm).

2.2. Chemicals and utilities

HCl (37%, 7647-01-0) from Merck & Co., and NaOH (≥98%, 1310-73-2) from Carl Roth GmbH were utilized to adjust of pH values during tests. H_2O_2 (35%, 7722-84-1) from Carl Roth GmbH and $FeSO_4 \cdot 7H_2O$ (99.5%, 7782-63-0) from Merck & Co. were utilized in the Fenton oxidation tests. The aerobic and anaerobic activated sludge samples were collected from the local municipal wastewater treatment plant in Schönerlinde, Berlin. The aerobic activated sludge, with a dry matter concentration of 6.7 g/L, was collected directly from the aeration pool. It was further used in the respiration inhibition assays. The anaerobic sludge was used for the biogas production tests. Table 2 gives an overview of the physico-chemical properties of the activated sludges.

Table 1 | Composition of wastewater

Components	Value
pH	8.9
Conductivity [mS/cm]	11.83
Total phenol index [g/L]	1.4
COD [g/L]	16.4
DOC [g/L]	5.3
NH_4^+-N [g/L]	1.5
NO_3^--N [mg/L]	67.5
TN [g/L]	2.3

Table 2 | Properties of the utilized activated sludge

Components	Aerobic sludge	Anaerobic sludge
Dry matter content [g/L]	6.7	24.5
Organic dry matter [g/L]	5.4	17.4
COD_{liquid} [mg/L]	97.5	523.6
DOC_{liquid} [mg/L]	30.3	171.6
Usage	Respiration inhibition assays	Biogas production

2.3. Experimental design

2.3.1. Fenton reaction

The Fenton experiments were conducted in batch reactors as shown in Figure 1 in supplemental material. Here, 100 mL of the wastewater sample was treated in quartz stube with or without UV irradiation for 120 min. Single-factor test series were conducted to isolate the effect of the initial pH value, Fe^{2+} and H_2O_2 dosage on their detoxification efficiency and economic viability.

To estimate the most efficient pH value of Fenton oxidation, the initial pH value was set at 3, 5 and unchanged condition of pH 8.9. The H_2O_2 with 30% of the normalized stoichiometric ratio (NSR) to COD (275 mM H_2O_2) and Fe^{2+} ions with a molar ratio of H_2O_2:Fe^{2+} at 15:1 (18.3 mM of Fe^{2+}) were dosed as oxidant and homogeneous catalyst in feed. Subsequently, the most cost-efficient Fe^{2+} dosage was determined from tests at initial pH 5 and an initial H_2O_2 dosage of 30% NSR, among molar ratios of H_2O_2:Fe^{2+} between 10 and 30. To estimate the optimal H_2O_2 dosage in range of 15–100% NSR, the tests with conditions of initial pH 5, molar ratio of H_2O_2:Fe^{2+} at 15:1 were investigated.

To estimate the chemical costs of Fenton oxidation, the samples were neutralized to pH 7 ± 0.5 by NaOH. The iron hydroxide was precipitated through coagulating or adsorbing with humic substances and other organic compounds. To estimate the sludge production and organic content, the entire precipitated sludge was filtered through a paper filter. The sludge production was quantified after overnight drying at 105 °C.

To prepare the Fenton oxidation treated samples for further analysis and experiments, 5 g/L MnO_2 were added into feed solution to remove residual oxidants. Due to adsorption effect of phenolic compounds onto MnO_2, the concentration of phenolic compounds was decreased by 34–38% in the blank control by this procedure. However, the concentration reduction of phenolic compounds by adsorption of MnO_2 was not subtracted.

2.3.2. Respiration inhibition assays

The respiration inhibition assays were conducted based on the modified OECD guideline 209. The experimental setup is depicted in Figure 2 in the supplementary material. They were conducted with a total residence time of 3 days at 25 °C.

The OxiTop®-C works at absolute pressure of $1,000 \pm 250$ mbar. As the whole air in reactor was blown out by pure oxygen, the maximal reduction of partial oxygen pressure in this respiration assay is 0.25 bar. The dissolved oxygen in the feed samples is considered within 25% deviation. The 0.25 bar partial oxygen pressure corresponds to a biochemical oxygen consumption of 357 mg. Thus, considering the biological availability, the COD of samples was set at 550 mg.

During respiration inhibition assays, 100 mL feed samples were prepared through mixing of wastewater samples (85% COD), synthetic sewage (15% COD) and desalinated water. These prepared samples were mixed after that with 100 mL of aerobic activated sludge in a 1 Liter three-neck laboratory bottle, subsequently to adjust the pH to 7 ± 0.2. Before the bottles were airtightly closed, the air in bottles was totally purged out with pure oxygen. During the respiration assays, the produced CO_2 was totally absorbed by NaOH platelets in the gas phase of the reactor. The pressure reduction in the bottle was congruent to the total oxygen consumption. To decrease the experimental error of pressure change from unstable temperatures, the respiration inhibition tests were stirred by magnetic stirrer under thermostable conditions. The composition of the feed samples is shown in Table 3. The O_2 consumption was calculated by the following Equation (1):

$$V_{O_2} = \Delta p \times V_{bottle} \tag{1}$$

where the V_{O_2} is the consumption of O_2 (mL), Δp is the pressure change during tests (bar), V_{bottle} is the gas volume of 800 mL in the bottle.

2.3.3. Biogas production test

To show the efficiency of the Fenton oxidation on wastewater detoxification, the original wastewater and the Fenton oxidation-treated wastewater were provided in 20 days' biogas production tests. Before tests, all samples were adjusted to pH 7 ± 0.3. The composition of samples is described in Table 4. The experimental setup is depicted in Figure 3 in the supplementary material.

2.4. Analytical methods

The total phenol concentration was detected through a wet chemical color-spectrophotometric analysis method using the Folin–Ciocalteau reagent. The method is described in the supplementary material. The concentration of NH_4^+ was detected by flow injection analysis (FIA) FOSS FIA Star 5000. The concentrations of dissolved organic carbon and total nitrogen in samples were measured by thermal catalytic oxidation process at 850 °C with an Analytik Jena multi N/C 3100 + CLD. The chemical oxygen demand (COD) was detected through total combustion of sample at 1,200 °C by COD analyzer from company LAR process Analyzers AG. The iron concentrations in wastewater samples were detected by atomic absorption spectrometry (AAS) with model 900AA from Perkin Elmer-AAS-PinAAcle™.

Table 3 | Composition of feed samples in respiration inhibition assays

Sample

Name		COD [g/L]	Volume [mL]	Total COD [mg]	V_{water} [mL]	V_{Sewage} [mL]	COD_{Sewage} [mg]	V_{Sludge} [mL]
Blank		0	0	0	96.3	3.7	100	100
Original		16.4	33	550	62.7	3.7	100	100
Optimal (5, 15, 30%)		8.7	63.5	550	32.8	3.7	100	100
pH	pH 3	8.7	63.2	550	33.1	3.7	100	100
	pH 8.9	14.5	37.9	550	58.4	3.7	100	100
UV-Fenton	pH 5	9.0	61.6	550	34.7	3.7	100	100
	pH 8.9	14.3	38.6	550	57.7	3.7	100	100
H_2O_2: Fe^{2+}	10	8.4	62	550	34.3	3.7	100	100
	20	8.9	62.3	550	34	3.7	100	100
	30	9.2	60	550	36.3	3.7	100	100
NSR of H_2O_2	15%	11.3	48.6	550	47.7	3.7	100	100
	45%	7.4	75	550	21.3	3.7	100	100
	60%	6.4	81.5	550	14.8	3.7	100	100
	100%	6.4	86.1	550	10.1	3.7	100	100

Table 4 | Samples for the biomethane assays

Sample	V$_{sample}$ [mL]	V$_{sludge}$ [mL]	V$_{water}$ [mL]	COD$_{begin}$ [mg]	
				COD$_{Sample}$	COD$_{Sludge}$
Original	75.2	100	44.8	1,151	168
Fenton	85.4	100	34.6	800	168
Blank	0	220	0	0	370

3. RESULTS AND DISCUSSION

3.1. Influence of pH, H_2O_2 dosage and Fe^{2+} dosage on degradation

The optimal pH was determined among the pH values 3, 5 and 8.9 (original pH). The tests were conducted with a 30% NSR H_2O_2 dosage (275 mM) and a molar ratio of H_2O_2:Fe^{2+} at 15:1 (18.3 mM Fe^{2+}). After 2 hours' reaction, the pH values of the samples decreased from 3 to 2.48, 5 to 2.69 and 8.9 to 7.93 (Figure 1(a)). Similar removal rates of phenolic compounds, COD and NH_4^+ were observed in tests with initial pH 3 and 5. The sample without pH adjustment (pH 8.9) showed a higher ammonium removal rate (30%). Two effects contributed to this phenomenon: a lower concentration of ammonium was generated from nitrogen-containing organic compounds by Fenton oxidation at pH 8.9; and a higher escape rate of ammonia was achieved at alkaline conditions of pH 8.9 (Gamaralalage *et al.* 2019).

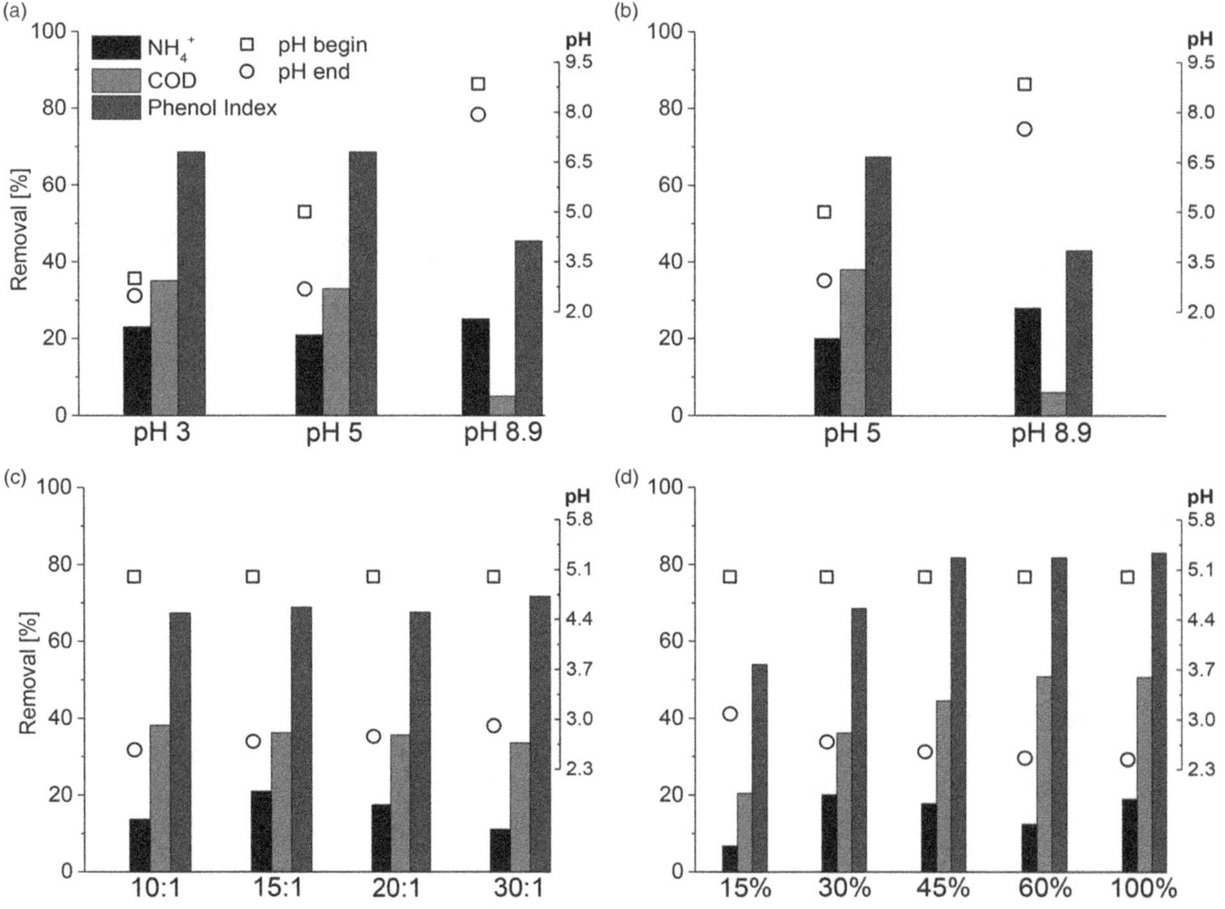

Figure 1 | The removal rates of ammonium, COD, phenol index as well as pH decrease after the Fenton reaction for 120 min with or without UV irradiation. (a) pH = 3, 5 and 8.9, Fe^{2+} = 18.3 mM, H_2O_2 = 275 mM; (b) UV = 15 W, pH = 5 and 8.9, Fe^{2+} = 18.3 mM, H_2O_2 = 275 mM; (c) pH = 5, H_2O_2 = 275 mM, molar ratio of H_2O_2:Fe^{2+} = 10–30; (d) pH = 5, molar ratio of H_2O_2 : Fe^{2+} = 15, H_2O_2 = 137.5–917 mM (NSR at 15–100%).

Due to the pH buffering system (such as humic substances), a further pH lowering from pH 5 to pH 3 was deemed uneconomical in this application scenario. Owing to an obvious higher COD removal rate in comparison with pH 8.9, the optimal pH value of the Fenton oxidation was determined to be pH 5. This contrasts with other research and the theoretical basics of Fenton oxidation, that the most effective pH of Fenton oxidation was found at around 2.5 (Pignatello *et al.* 2006). The aqueous phase from hydrothermal carbonization process contains high concentrations of humic substances (Usman *et al.* 2020). Humic substances building complex bonds to ionic iron, act as chelating agents in Fenton oxidation processes. Through applying chelating agents in feed, the Fenton oxidation achieved also high efficiency at pH 5–8 (Zhang & Zhou 2019).

UV irradiation can regenerate Fe^{2+} from Fe^{3+}, and improve the COD degradation efficiency in Fenton experiments (Hu *et al.* 2011; Nidheesh *et al.* 2021). A 15 W UV lamp was used to irradiate the wastewater in tests of initial pH 5 and 8.9. However, in the comparison in Figure 1(a), the removal rates of COD and ammonium were not obviously improved as shown in Figure 1(b). The COD removal rate was slightly increased from 33% to 38% under UV irradiation in the test with an initial pH of 5. The wastewater in this work had an intense color, likely impeding the transfer of UV light. Compared with the initial Fe^{2+} concentrations in a referenced publication (40 mg/L), it is much higher (1,021 mg/L) in this work. Thus, the COD degradation was not obviously enhanced by regenerated Fe^{2+}.

The iron and H_2O_2 dosages are the two main operational parameters in Fenton oxidation. For the treatment of similar industrial wastewater (originating from e.g. petroleum refineries, pharmaceutical and antibiotic fermentation processes) the molar ratios of H_2O_2:Fe^{2+} have been set in range of 8.2–20 (Ribeiro & Nunes 2021). Based on 275 mM H_2O_2 dosage (NSR value of 30%), the molar ratio of H_2O_2:Fe^{2+} was set between 10 and 30 in this work. At these reaction conditions, the Fe^{2+} concentration showed little influence on the removal rates of phenolic compounds and COD. Their removal rates ranged at 67–72% and 34–38%, respectively. The pH value of the feed was more obviously decreased at higher Fe^{2+} dosage. With a molar ratio of H_2O_2:Fe^{2+} of 10, 15, 20, 30, the pH values decreased from 5 to 2.57, 2.69, 2.76 and 2.91, respectively. Among these tests, the highest NH_4^+ removal (22%) was achieved at a molar ratio of 15:1.

To investigate the correlation between H_2O_2 dosage and the degradation efficiency, H_2O_2 was dosed between 15 and 100% NSR with a molar ratio to Fe^{2+} of 15:1. The drop in pH values after Fenton oxidation was increased by higher H_2O_2 dosage, correspondingly to final pH values between 3.08 and 2.44 (Figure 1(d)). The phenolic compounds and COD were degraded more efficiently by higher H_2O_2 dosage. After increasing the H_2O_2 dosage from 15% to 45% NSR, their removal rates increased from 54% to 82% and 15% to 45%. The removal rate of COD plateaued at 30% NSR and reached a maximum of 51% at H_2O_2 dosage of 60% NSR. The removal efficiency of COD by Fenton reaction was efficiently improved through a multistage neutralization–coagulation process with limestone (Guo *et al.* 2018). However, the precipitates must be treated as hazardous solid waste. Due to the increased sludge production yield from the lime and its secondary effect to organic compounds such as adsorption or agglomeration, larger amounts of sludge should be managed (Gao *et al.* 2022). The solid waste management requirements limit the applicability of Fenton oxidation in industrial-scale utilization.

The NH_4^+ removal rates were comparably lower, ranging between 12 and 20% with H_2O_2 dosages of higher than 30% NSR. The nitrogen-containing organic compounds shared around one-third of the nitrogen-source in the condensate, this is connected to the formation of ammonium as a by-product from the decomposition of nitrogen-containing organic compounds. The same phenomenon was observed in palm oil mill effluent treatment, a removal rate of NH_4^+ was achieved by 97% within 15 min, and subsequently decreased to 35% after 90 min reaction time (Gamaralalage *et al.* 2019).

This work focused on the detoxification of wastewater by Fenton oxidation, in comparison with the mentioned paper, the dosages of H_2O_2 and Fe^{2+} were reduced by 30 and 93%. (Gamaralalage *et al.* 2019). For a higher removal rate of NH_4^+, the electro-Fenton oxidation could be performed. As announced by Menon and coworkers, the NH_4^+ removal rate reached 59% in textile wastewater treatment (Menon *et al.* 2021).

3.2. Iron precipitation

To make a complete economic evaluation of pretreatment by Fenton oxidation, the chemical consumption and sludge production yield after neutralization of NaOH are summarized in Table 5. To further provide a sustainable reuse application of precipitates, the iron content in sludge was also determined.

With an initial dosage of H_2O_2 and Fe^{2+} at 30% NSR and 1,021 mg/L, the amount of precipitated sludge after neutralization reached 7.5–9.5 kg/ton wastewater in Figure 2. The iron content reached 9.9% in the sludge sample of pH 8.9, which was much higher than the content in the samples of pH 3 or pH 5. This is a consequence of a lower solubility of iron hydroxide at neutral pH value, transferring most iron into the solid sludge. Su and coworkers researched the influence of the final

Table 5 | Mass balance of Fe after Fenton reaction

Sample		pH	NSR H$_2$O$_2$ [%]	Fe$^{2+}_{input}$ [mg/L]	Fe$_{Solution}$ [mg/L]	$W_{Fe_{solution}}$ [%]	Sludge [kg/ton]	Fe in sludge [m.-%]
pH	pH 3	3	30	1,021	601	59	9.5	4.3
	pH 5	5	30	1,021	595	58	7.5	4.9
	pH 8.9	8.9	30	1,021	21	2	9.4	9.9
NSR of H$_2$O$_2$	15%	5	15	511	309	60	6.4	3
	30%	5	30	1,021	713	70	7.5	4
	45%	5	45	1,531	187	12	10	12.5
	60%	5	60	2,042	599	29	11.2	12
	100%	5	100	3,404	83	2	18.3	18.8
H$_2$O$_2$:Fe	10	5	30	1,532	384	25	10.6	10.1
	15	5	30	1,021	713	70	7.5	1.0
	20	5	30	766	591	77	8	2.2
	30	5	30	511	494	97	6.6	0.4

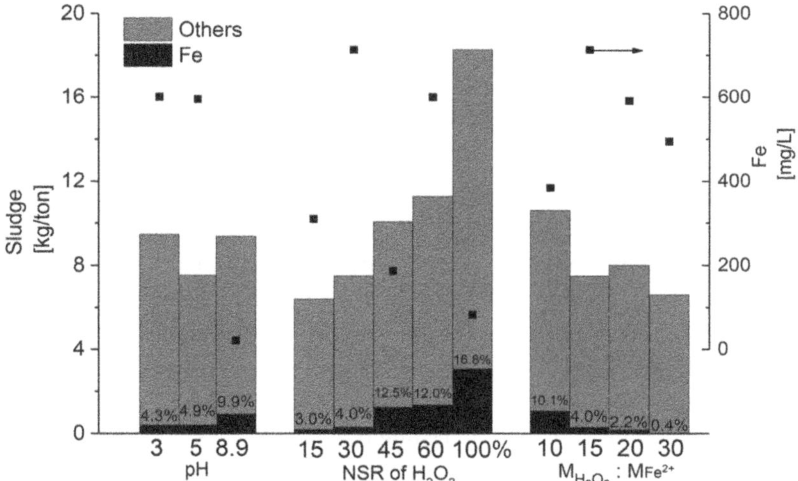

Figure 2 | The iron concentration in treated wastewater and its content in generated sludge after the Fenton reaction.

pH on the sludge production. With final pH of 3, the sludge production yield was higher than that at pH 5 (Su *et al.* 2019). However, the sludge in this work was collected after neutralization, the lower sludge production from test with initial pH 5 might be owing to a slight alkalization in neutralization step.

The sludge production yield and iron content of the sludge augmented along with the H$_2$O$_2$ and Fe^{2+} concentration in the sample. With a molar ratio of H$_2$O$_2$:Fe^{2+} at 15:1, the H$_2$O$_2$ was dosed in the range of 15–100% NSR into the feed, increasing the sludge yield from 6.4 to 18.3 kg sludge/ton wastewater. The content of iron grew correspondingly from 3 to 16.8%. The highest sludge production yield reached 18.3 kg/m^3 with an iron content of 16.8%, in tests with H$_2$O$_2$ dosage of 100% NSR. Under the operational condition of pH 5, the molar ratio of H$_2$O$_2$:Fe^{2+} of 15 and H$_2$O$_2$ dosage of 30% NSR, the sludge production reached 7.5 kg/ton with an iron content of 4.5%.

In comparison to initial wastewater, the content of toxic compounds such as polycyclic aromatic hydrocarbons is lower in Fenton sludge (Wang *et al.* 2019). The Fenton sludge could be processed in anaerobic digestion trials for biogas production. An enhancement on hydrolysis of macromolecular organic compounds was also found in the hydrolysis or acidification process of wastewater treatment (Wang *et al.* 2022). Moreover, it was applied as source material for the production of magnetic adsorbent or catalysts (Liu *et al.* 2017; Gan *et al.* 2020; Tong *et al.* 2021). In this work, the Fenton sludge could be introduced as feedstock into a gasifier. Compared with other publications, the Fe^{2+} dosage is obviously lower in this work, the ash production of the gasifier will not be seriously impacted by the introduced Fenton sludge. (Gamaralalage *et al.* 2019; Wu *et al.* 2021).

In the experiment with 100% NSR H_2O_2 dosage, the Fe^{2+} concentration was set at 3,404 mg/L. The residual iron concentration in this sample was low at 83 mg/L. This corresponds to a 98 wt.-% transfer of iron into the sludge. One reason for the near complete precipitation of the iron is the overdosed H_2O_2 at 100% NSR. The generated hydroxyl radicals oxidized the Fe^{2+} to Fe^{3+}, wasting their oxidation potential on the iron. (Nidheesh *et al.* 2021). Due to the low solubility of Fe^{3+} hydroxides, the iron was more efficiently filtrated from feed, resulting in a low iron ion concentration after neutralization (Cravotta 2008).

In the other samples, the iron ion concentrations kept between 187 and 713 mg/L. As shown in previous publications, the residual iron ions negatively impacted the nitrogen removal in subsequent biological treatments (Philips 2003). With an iron ion concentration of 112 mg/L (Fe^{2+} or Fe^{3+}), the nitrogen removal was 40% or 60% lower than the iron free blank sample. After adaptation for 12 days, they were still 9% or 30% lower. Thus, the wastewater after Fenton oxidation may inhibit on nitrification process.

3.3 Respiration inhibition assays

The total oxygen consumption of samples are shown in Figure 3. The blank sample containing only activated sludge and synthetic sewage, showed a sharp increase in oxygen consumption within the first 8 hours and a flat curve that was moderately constant over the remaining 64 hours in Figure 3(a). The highest oxygen consumption rate reached 7.2 mL/h in 5 hours. The total oxygen consumption was accounted for at 100 mL. Compared with the blank sample, the original sample showed a slightly higher oxygen consumption rate of 7.5 mL/h at 7 h. Owing to toxic compounds in the wastewater, the highest oxygen consumption rate was 2 hours delayed, the total oxygen consumption reached as low as 56.8 mL in 72 hours.

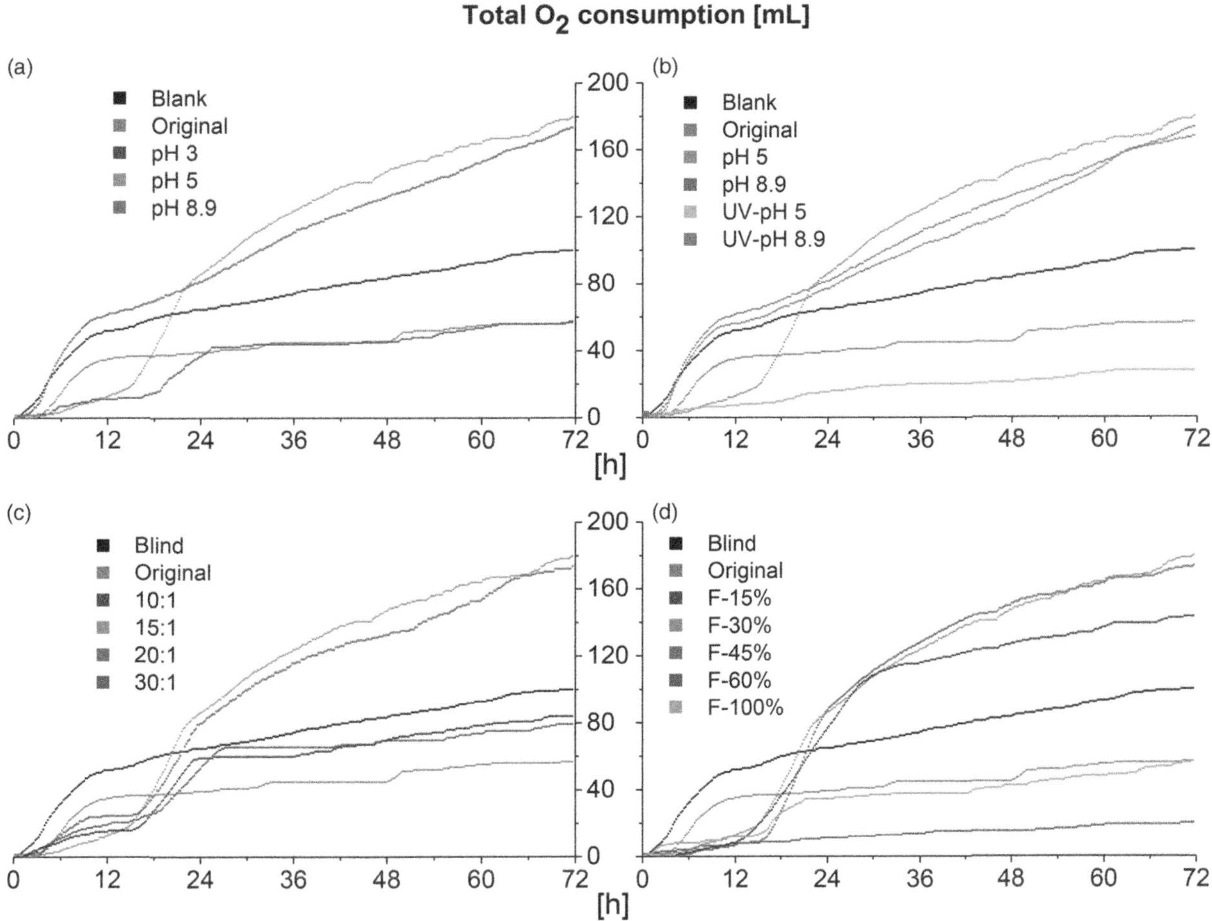

Figure 3 | Total oxygen consumption of feed samples during respiration inhibition assays. (a) Influence of initial pH on toxicity; (b) influence of UV irradiation on toxicity; (c) influence of molar ratio of H_2O_2:Fe on toxicity; (d) influence of H_2O_2 dosage on toxicity.

After Fenton oxidation with an initial pH of 3 or 5, the toxic transformed organic compounds might be built in feed. At the beginning of the respiration assays, the oxygen consumption of samples at pH 3 and pH 5 had underperformed compared with the original sample (Figure 3(a)). Their oxygen consumption rates reached the highest value of 5.6 and 10.4 mL/h at residence times of 20 and 21 h. Compared with the sample at pH 3 (lag phase with 18 h), less amounts of toxic compounds were generated at pH 5 (lag phase with 14 h).

The toxic organic compounds were efficiently removed after the Fenton reaction with an initial pH of 8.9, the lag phase of the sample at pH 8.9 was comparable to the blank sample, its highest oxygen consumption rate reached 11.2 mL/h in 5 hours. The total oxygen consumption reached among tests a highest value of 180 mL, in the sample at pH 5. Considering the COD removal rate and total oxygen consumption, the optimal pH value was indicated at pH 5.

Figure 3(b) shows a similar oxygen (Table 6) for the wastewater samples at pH 8.9 with and without UV irradiation. However, the oxygen consumption rate of the sample UV-5 is much lower than for the sample without UV at pH 5. Although a slightly higher COD removal rate was achieved by UV-Fenton oxidation, larger amounts of transformed toxic compounds might be generated at pH 5. Thus, considering the increased toxicity, UV irradiation is not recommended.

The influence of initial Fe^{2+} concentration on the toxicity of treated wastewater was also investigated. The Fe^{2+} was dosed into the wastewater with a molar ratio of H_2O_2:Fe^{2+} in the range of 10–30 that corresponds to an initial Fe^{2+} concentration of 1,532–511 mg/L. As shown in Figure 3(c), the wastewater sample of 15:1 and 20:1 showed a higher total oxygen consumption than the original sample for the residence time of 16 h. With the 72 h residence time, the total oxygen consumption reached 173.6 (20:1), 176.6 (15:1), 79.2 (10:1) and 84 mL (30:1). This indicates an optimal molar ratio of H_2O_2:Fe^{2+} at 15:1 in Fenton oxidation.

To determine the influence of the hydrogen peroxide dosage on the toxicity of Fenton oxidation-treated wastewater, the tests were conducted with a fixed molar ratio of H_2O_2:Fe^{2+} at 15:1 and a varied H_2O_2 dosage of 15–100% NSR. The sample F-15% shows the lowest oxygen consumption rate in Figure 3(d). With the residence time of 72 h, the total oxygen consumption reached 20 mL. The toxicity of wastewater was decreased with a higher H_2O_2 dosage. With the initial H_2O_2 dosage of 30 and 45%, the oxygen consumption rates reached their highest values of 10.4 and 11.2 mL/h at the residence time of 21 h. However, the biological available organic compounds were degraded by highly dosed H_2O_2, the highest oxygen consumption rates of samples F-60% and F-100% reached lower values of 7.2 and 5.6 mL/h.

As identical results, the samples of F-30% and F-45% observed total oxygen consumption of 180 and 173 mL at the residence time of 72 h. Owing to lower chemical consumption, the favorite H_2O_2 dosage was estimated at 30% NSR. Thus, through the respiration assays, the optimal condition of the Fenton oxidation was determined as pH 5, with the H_2O_2 dosage of 30% NSR and a molar ratio of H_2O_2:Fe^{2+} at 15:1.

3.4. Biogas production

The biogas production of Fenton sample (F515), original wastewater and anaerobic sludge over a period of 20 days is shown in Figure 4. The original wastewater displayed a noteworthy inhibition on biogas production. Compared with the blank or the

Table 6 | Highest oxygen consumption rate and retention time

Sample		$r_{highest}$ [mL/h]	$t_{highest}$ [h]
Blank		7.2	5
Original		7.5	7
Optimal (5, 15, 30%)		10.4	21
pH	pH 3	5.6	20
	pH 8.9	11.2	5
UV-Fenton	pH 5	1.6	–
	pH 8.9	11.2	4
H_2O_2: Fe^{2+}	10	4.8	20
	20	8.8	22
	30	6.4	18
NSR of H_2O_2	15%	1.6	–
	45%	11.2	22
	60%	7.2	21
	100%	5.6	18

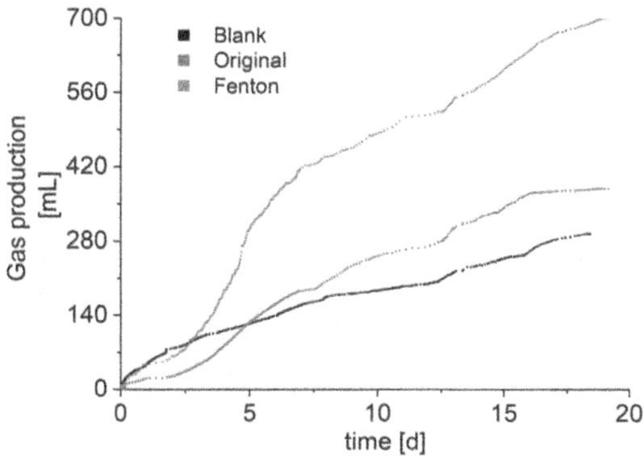

Figure 4 | Biogas production of original sample, sample after the Fenton reaction and anaerobic sludge.

treated sample, only roughly one-third biogas was produced in first 3 days. The anaerobic sludge had the lowest absolute gas production, oppositely a highest biogas production relating to the COD removal. This implies the decay of organisms at nutrient lack conditions, the released nutrition from organisms were utilized as feedstock of biogas production.

The sample of Fenton produced 84% more biogas than the original sample in around 20 days. That confirms the 66% higher gas production relating to COD removal. A higher concentration of biodegradable organic compounds, especial the volatile fatty acids was generated after the Fenton reaction (Feki *et al.* 2020). Through detection by Hach-Lang cuvette LCK365, the organic acid concentration in the wastewater was increased from 1.02 to 2.03 g butyric acid equivalents (BAE)/L after Fenton oxidation. After anaerobic digestion for around 20 days, the concentration of organic acid was decreased to 0.28 g BAE/L.

The LCK 345 cuvette from the Hach-Lang company was developed based on the 4-aminoantipyrine (AAP) method to measure the concentration of small molecular phenolic compounds. It was utilized to track the small phenolic compound concentrations over the wastewater treatment processes. After pretreatment by Fenton oxidation, the concentration of small molecular phenolic compounds was decreased by 70%, from 184 to 55 mg/L. It was further decreased to 24 mg/L after 20 days' anaerobic treatment. In contrast, 84 mg/L of small molecular phenolic compounds were found in the original wastewater after anaerobic treatment. The organisms had an inhibitory effect from phenol with an aqueous concentration of 90 mg/L (Zhao *et al.* 2015). Therefore, the lower biogas production yield in the original sample might be owing to a continuous inhibition of small molecular phenolic compounds to organisms (Table 7).

Figure 5 shows the mass flow of COD over the course of the Fenton oxidation and anaerobic treatment. Through Fenton oxidation, the COD concentration was decreased from 16.4 to 8.7 g/L, equivalent to a 53% COD removal. During anaerobic treatment, the COD concentration was further diluted by the mixed anaerobic sludge. Over the digestion for 20 days, the COD concentration was further decreased from 4.43 to 2.41 g/L. The COD concentration could be further decreased with a longer digestion time. For future work, the adsorption effects of anaerobic sludge, biomass accumulation and the composition of produced biogas could be investigated.

3.5. Cost

The cost of Fenton oxidation was calculated from chemical consumption, sludge disposal, electricity cost and the operation costs. The energy cost of drying or transport were not focused on. To be informed by statistics from the Independent Commodity Intelligence Services (ICIS), the market price of technical grade HCl in Germany was assessed at 35–70 €/ton in

Table 7 | Change of wastewater parameters over 20 days' anaerobic treatment

Sample	COD$_{removal}$ [%]	Gas production yield [mL]	Gas production yield relating to COD removal [mL/mg]
Original	30	380	0.97
Fenton	45	700	1.61
Anaerobic sludge	20	293	2.92

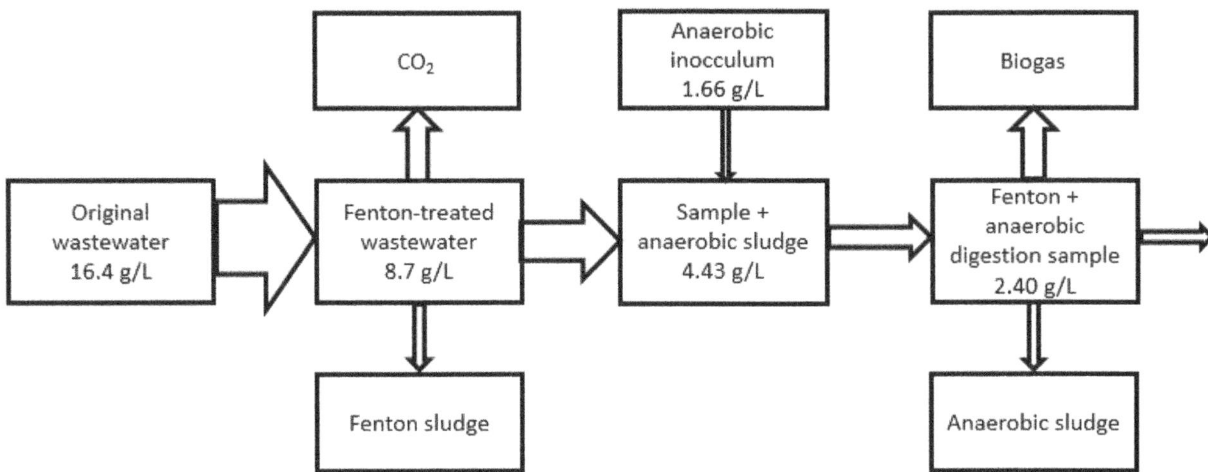

Figure 5 | COD mass flow diagram of Fenton and anaerobic treatment.

2019, thus the price of HCl at 60 €/ton was used in this work (Barker 2019). The sale price of H_2O_2 varied in the range 700–1,200 $/ton, in this work it was fixed at 800 €/ton (Ciriminna *et al.* 2016). The price of ferrous sulfate heptahydrate was estimated at 115 $/ton (97 €/ton) (Fac. MR 2018) (Table 8).

The NaOH was utilized to neutralize the pH value of treated wastewater to around 7. The price of NaOH was indicated with 300–500 €/ton (400 €/ton in this work) (Barker 2018). The sludge disposal costs varied strongly based on geographical region and composition. In the region of Lahn-Dill in Germany, the disposal price of the industrial sludge, which may have a high iron content was set at 220 €/ton (Lahn-Dill 2018). In this work, the disposal price of sludge was estimated at 220 €/ton. The electric consumption of the Fenton process was largely independent of the operational parameters. The energy consumption of UV irradiation was defined as additional electricity in Table 9, the price of electricity for industrial utilization was set at 0.11 €/kwh (Hein *et al.* 2021).

The operational costs of Fenton oxidation with different setting conditions are shown in Figure 6. The absolute costs of Fenton oxidation were between 7.3 and 33.5 euro per ton wastewater, while the specific COD elimination costs varied between 2.2 and 18.8 €/kg COD_{eli}. With an initial pH of 3 or 5, the main costs were the hydrogen peroxide with a share of 56–59%, this was used for neutralization (18%) and the sludge disposal observed a share of 13–16%. The sample at pH 8.9 showed the least absolute cost (no chemical costs for the adjustment of the pH value before and after the process), however owing to an inefficient reaction, it performed the high relative cost of 12.2 €/kg COD_{eli}.

The electricity cost of 2 hours' UV irradiation was at around 8.3 €/ton wastewater. Although the electricity cost shared around 40–46% of total costs in UV-Fenton oxidation, the organic compounds were not obviously better degraded. The cost of COD degradation reached 3.46 and 18.83 €/kg COD_{eli} in tests with an initial pH of 5 and 8.9. Considering the high energy cost, the UV-Fenton oxidation is not recommended in the treatment of the condensate from gasification processes.

Table 8 | Cost items

Consumption	Price
HCl [€/ton]	60
NaOH [€/ton]	400
H_2O_2 [€/ton]	800
$FeSO_4{\cdot}7H_2O$ [€/ton]	97
Disposal of sludge [€/ton]	220
Electricity cost [€/kwh]	0.11

Table 9 | Parameter costs of Fenton oxidation or UV-Fenton oxidation in laboratory scale

Sample		Acid [€/ton]	H_2O_2 [€/ton]	$FeSO_4\cdot7H_2O$ [€/ton]	Neutralization [€/ton]	Sludge [€/ton]	Additional electricity [€/ton]	Total [€/ton]	COD removal [%]	Relative cost [€/kg COD]
pH	3	0.87	7.46	0.5	2.3	2.1	–	13.3	35	2.4
	5	0.75	7.46	0.5	2.2	1.7	–	12.6	35	2.4
	8.9	0	7.46	0.5	0.1	2.1	–	10.1	5	12.2
UV	5	1.25	7.46	0.5	1.83	2.0	8.3	20.8	38	3.5
	8.9	0	7.46	0.5	0.12	1.5	8.3	17.9	6	18.8
NSR of H_2O_2	15%	0.75	3.73	0.25	1.1	1.4	–	7.3	20	2.3
	30%	0.75	7.5	0.5	2.2	1.7	–	12.6	35	2.2
	45%	0.75	11.2	0.75	2.7	2.2	–	17.6	45	2.5
	60%	0.75	14.9	1	3.1	2.5	–	22.3	51	2.8
	100%	0.75	24.9	1.67	2.2	4.0	–	33.5	51	4.2
H_2O_2: Fe	10	0.75	7.5	0.75	2.3	2.3	–	13.6	38	2.3
	15	0.75	7.5	0.5	2.2	1.7	–	12.6	35	2.2
	20	0.75	7.5	0.38	1.9	1.8	–	12.2	35	2.2
	30	0.75	7.5	0.25	1.7	1.5	–	11.6	34	2.2

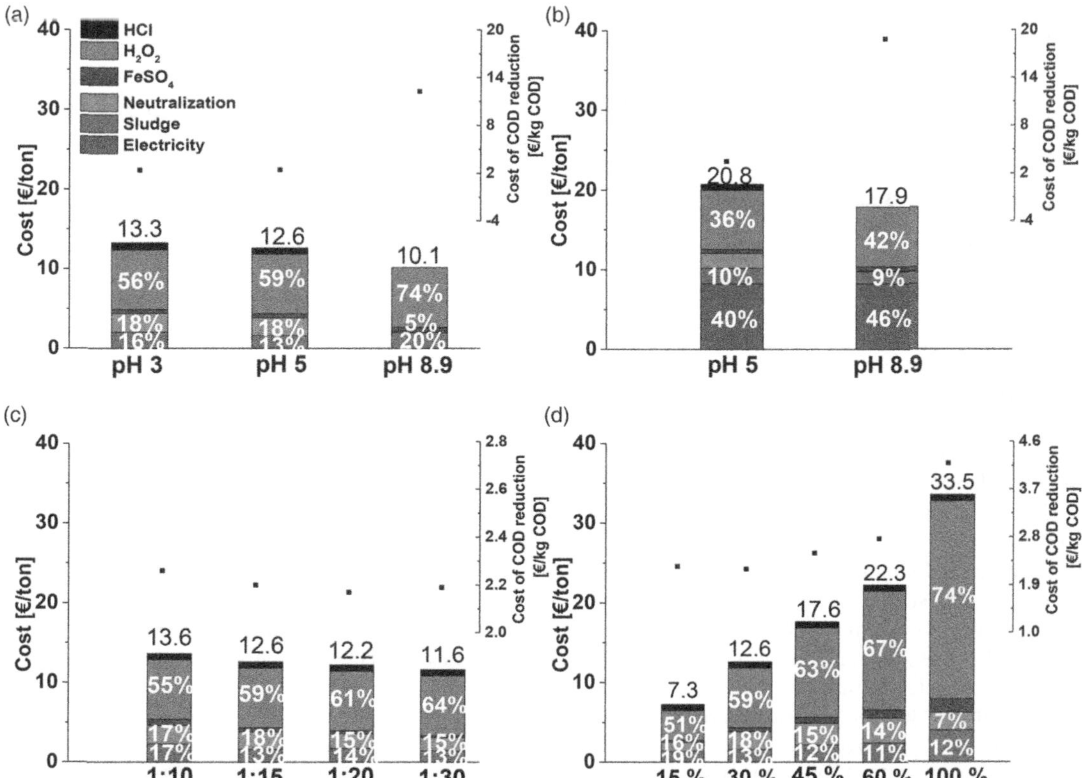

Figure 6 | Total cost and cost of COD reduction after Fenton reaction or UV-H_2O_2 oxidation with variate conditions. (a) pH = 3, 5, 8.9, Fe^{2+} = 18.3 mM, H_2O_2 = 275 mM; (b) UV = 15 W, pH = 5 and 8.9, Fe^{2+} = 18.3 mM, H_2O_2 = 275 mM; (c) pH = 5, H_2O_2 = 275 mM, molar ratio of H_2O_2: Fe^{2+} = 10–30; (d) pH = 5, molar ratio of H_2O_2:Fe^{2+} = 15, H_2O_2 = 137.5–917 mM (NSR value of 15–100%).

The influence of the initial Fe^{2+} on the costs was also investigated. As shown in Figure 6(c), the three highest costs of these tests were H_2O_2 consumption (55–64%), neutralization (15–18%) and sludge disposal (13–27%). Owing to a higher Fe^{2+} dosage, the costs of neutralization, sludge disposal in the test with a molar ratio of H_2O_2 : Fe^{2+} of 10 : 1, was slightly higher than others. However, owing to a higher COD removal efficiency, the relative costs of COD elimination were similar in the range 2.26–2.19 €/kg COD_{eli}.

Figure 6(d) shows the cost of Fenton oxidation with the H_2O_2 dosage of 15–100% NSR at molar ratio for H_2O_2:Fe^{2+} of 15:1. The largest share of expenses in these tests is the H_2O_2. With rising H_2O_2 dosages from 15% to 100% NSR, the cost share of H_2O_2 increases from 51% to 74%. The COD removal increased correspondingly from 20% to 51%. That conforms to the increasing relative cost of COD elimination from 2.2 to 4.2 €/kg COD_{eli}. Moreover, the costs of sludge disposal were increased correspondingly from 1.4 to 4.0 €/ton. The lowest cost of COD degradation was 2.2 €/kg at the H_2O_2 dosage of 30% NSR.

4. CONCLUSIONS

The detoxifying effect of Fenton oxidation on a biomass gasification condensate was assessed using respiration inhibition assays. Through the thorough analysis of the samples after Fenton oxidation, the most cost-efficient operational parameters were determined with a total operational cost of 2.2 €/kg COD_{eli}. For the test with an initial pH 5, 30% NSR dosage of hydrogen peroxide and a molar ratio of H_2O_2:Fe^{2+} of 15:1 produced the most efficient pretreatment results. Compared with the original wastewater, the total oxygen consumption of the Fenton oxidation-treated sample was increased by 316% in a 3 days' respiration inhibition assay. The biogas production yield relating to COD removal was increased by 81% from 0.97 to 1.61 L_{biogas}/g COD_{eli}. Compared with the conventional Fenton reaction, which aimed at the complete COD removal, the dosage of H_2O_2 and Fe^{2+} was decreased by 30 and 93% in this work. Moreover, to sustainably reuse the Fenton sludge in the future, the sludge production yield and iron content in produced precipitates were found to be between 6.2–18.1 kg/ton wastewater and 0.4–16.8%.

However, two major issues of Fenton oxidation should be overcome: the low removal efficiency of ammonium and high residual iron concentration in the effluent. The electro-Fenton oxidation approaches aim to overcome these challenges. It requires lower consumption of oxidant and Fe^{2+}, leading to lower residual iron concentrations in effluent, higher removal rates of nitrogen and lower sludge yields.

DATA AVAILABILITY STATEMENT

All relevant data are included in the paper or its Supplementary Information.

REFERENCES

Babu, D. S., Srivastava, V., Nidheesh, P. V. & Kumar, M. S. 2019 Detoxification of water and wastewater by advanced oxidation processes. *Science of The Total Environment* **696**, 133961. Available from: https://linkinghub.elsevier.com/retrieve/pii/S0048969719339312 (accessed 13 August 2021).

Barker, C. 2018 *Chemical Profile: Europe Caustic Soda*. Available from: https://www.icis.com/explore/resources/news/2018/11/01/10276478/chemical-profile-europe-caustic-soda/

Barker, C. 2019 *German, Belgian HCl Contract Prices Rise for 2019*. Available from: https://www.icis.com/explore/resources/news/2019/01/25/10311220/german-belgian-hcl-contract-prices-rise-for-2019 (accessed 20 September2021).

Brillas, E. 2020 A review on the photoelectro-Fenton process as efficient electrochemical advanced oxidation for wastewater remediation. treatment with UV light, sunlight, and coupling with conventional and other photo-assisted advanced technologies. *Chemosphere* **250**, 126198. Available from: https://linkinghub.elsevier.com/retrieve/pii/S004565352030391X (accessed 3 February 2022).

Cetinkaya, S. G., Morcali, M. H., Akarsu, S., Ziba, C. A. & Dolaz, M. 2018 Comparison of classic Fenton with ultrasound Fenton processes on industrial textile wastewater. *Sustainable Environment Research* **28** (4), 165–170. Available from: https://linkinghub.elsevier.com/retrieve/pii/S2468203917303138 (accessed 3 February 2022).

Ciriminna, R., Albanese, L., Meneguzzo, F. & Pagliaro, M. 2016 Hydrogen peroxide: a key chemical for today's sustainable development. *ChemSusChem* **9** (24), 3374–3381. Available from: https://onlinelibrary.wiley.com/doi/10.1002/cssc.201600895 (accessed 20 September 2021).

Cravotta, C. A. 2008 Dissolved metals and associated constituents in abandoned coal-mine discharges, Pennsylvania, USA. Part 2: geochemical controls on constituent concentrations. *Applied Geochemistry* **23** (2), 203–226. Available from: https://linkinghub.elsevier.com/retrieve/pii/S0883292707002788 (accessed 28 January 2022).

Fact.MR 2018 *Ferrous Sulfate Market Forecast, Trend Analysis & Competition Tracking – Global Market Insights 2018 to 2028*. Available from: https://www.factmr.com/report/1954/ferrous-sulfate-market (accessed 20 September 2021).

Farzad, S., Mandegari, M. A. & Görgens, J. F. 2016 A critical review on biomass gasification, co-gasification, and their environmental assessments. *Biofuel Research Journal* **3** (4), 483–495. Available from: http://www.biofueljournal.com/article_32132.html (accessed 19 November 2021).

Feki, E., Battimelli, A., Sayadi, S., Dhouib, A. & Khoufi, S. 2020 High-rate anaerobic digestion of waste activated sludge by integration of electro-fenton process. *Molecules* **25** (3), 626. Available from: https://www.mdpi.com/1420-3049/25/3/626 (accessed 30 January 2022).

Gamaralalage, D., Sawai, O. & Nunoura, T. 2019 Degradation behavior of palm oil mill effluent in Fenton oxidation. *Journal of Hazardous Materials* **364**, 791–799. Available from: https://linkinghub.elsevier.com/retrieve/pii/S0304389418305405 (accessed 29 January 2022).

Gan, Q., Hou, H., Liang, S., Qiu, J., Tao, S., Yang, L., Yu, W., Xiao, K., Liu, B., Hu, J., Wang, Y. & Yang, J. 2020 Sludge-derived biochar with multivalent iron as an efficient Fenton catalyst for degradation of 4-chlorophenol. *Science of The Total Environment* **725**, 138299. Available from: https://linkinghub.elsevier.com/retrieve/pii/S004896972031812X (accessed 31 January 2022).

Gao, L., Cao, Y., Wang, L. & Li, S. 2022 A review on sustainable reuse applications of Fenton sludge during wastewater treatment. *Frontiers of Environmental Science & Engineering* **16** (6), 77. Available from: https://link.springer.com/10.1007/s11783-021-1511-6 (accessed 28 January 2022).

Grimm, D. & Wösten, H. A. B. 2018 Mushroom cultivation in the circular economy. *Applied Microbiology and Biotechnology* **102** (18), 7795–7803. Available from: http://link.springer.com/10.1007/s00253-018-9226-8 (accessed 10 August 2021).

Guo, Y., Xue, Q., Zhang, H., Wang, N., Chang, S., Fang, Y., Wang, H., Yuan, F., Pang, H. & Chen, H. 2018 Highly efficient treatment of real benzene dye intermediate wastewater by simple limestone and lime neutralization-coagulation with improved Fenton oxidation. *Environmental Science and Pollution Research* **25** (31), 31125–31135. Available from: http://link.springer.com/10.1007/s11356-018-3101-0 (Accessed 28 January 2022).

Hein, F., Herreiner, J., Graichen, D. P. & Lenck 2021 *Die Energiewende im Corona-Jahr: Stand der Dinge 2020*. Available from: https://www.agora-energiewende.de/veroeffentlichungen/die-energiewende-im-corona-jahr-stand-der-dinge-2020/

Hu, X., Wang, X., Ban, Y. & Ren, B. 2011 A comparative study of UV–Fenton, UV–H_2O_2 and Fenton reaction treatment of landfill leachate. *Environmental Technology* **32** (9), 945–951. Available from: http://www.tandfonline.com/doi/abs/10.1080/09593330.2010.521953 (accessed 27 September 2021).

Huang, J., Liu, J., Chen, J., Xie, W., Kuo, J., Lu, X., Chang, K., Wen, S., Sun, G., Cai, H., Buyukada, M. & Evrendilek, F. 2018 Combustion behaviors of spent mushroom substrate using TG-MS and TG FTIR: thermal conversion, kinetic, thermodynamic and emission analyses. *Bioresource Technology* **266**, 389–397. Available from: https://linkinghub.elsevier.com/retrieve/pii/S096085241830885X (accessed 10 August 2021).

Huang, J., Zhang, J., Liu, J., Xie, W., Kuo, J., Chang, K., Buyukada, M., Evrendilek, F. & Sun, S. 2019 Thermal conversion behaviors and products of spent mushroom substrate in CO_2 and N_2 atmospheres: kinetic, thermodynamic, TG and Py-GC/MS analyses. *Journal of Analytical and Applied Pyrolysis* **139**, 177–186. Available from: https://linkinghub.elsevier.com/retrieve/pii/S0165237018310003 (accessed 11 August 2021).

Ji, Q., Tabassum, S., Hena, S., Silva, C. G., Yu, G. & Zhang, Z. 2016 A review on the coal gasification wastewater treatment technologies: past, present and future outlook. *Journal of Cleaner Production* **126**, 38–55. Available from: https://linkinghub.elsevier.com/retrieve/pii/S095965261630066X (accessed 11 August 2021).

Kaur, P., Sangal, V. K. & Kushwaha, J. P. 2019 Parametric study of electro-Fenton treatment for real textile wastewater, disposal study and its cost analysis. *International Journal of Environmental Science and Technology* **16** (2), 801–810. Available from: http://link.springer.com/10.1007/s13762-018-1696-9 (accessed 3 February 2022).

Lahn-Dill 2018 *A. Preisliste (Direktanlieferung AWZ Aßlar). Abfallwirtschaft Lahn-Dill*. Available from: https://www.awld.de/de/Gebuehren-Preise/Preisliste/ (accessed 20 September 2021).

Liu, M., Xu, G. & Li, G. 2011 Gasification as a new thermal processes of sewage sludge utilization. In: *2011 International Conference on Multimedia Technology*, Hangzhou, China. IEEE, pp. 4454–4457. Available from: http://ieeexplore.ieee.org/document/6003362/ (accessed 19 November 2021).

Liu, U., Ou, C., Faheem, Shen, J., Yu, H., Jiao, Z., Han, W., Sun, X., Li, J. & Wang, L. 2017 Reuse of Fenton sludge as an iron source for $NiFe_2O_4$ synthesis and its application in the Fenton-based process. *Journal of Environmental Sciences* **53**, 1–8. Available from: https://linkinghub.elsevier.com/retrieve/pii/S1001074216301619 (accessed 31 January 2022).

Menon, P., Anantha Singh, T. S., Pani, N. & Nidheesh, P. V. 2021 Electro-Fenton assisted sonication for removal of ammoniacal nitrogen and organic matter from dye intermediate industrial wastewater. *Chemosphere* **269**, 128739. Available from: https://linkinghub.elsevier.com/retrieve/pii/S0045653520329374 (accessed 5 February 2022).

Mishra, L., Paul, K. K. & Jena, S. (2021) Coke wastewater treatment methods: mini review. *Journal of the Indian Chemical Society* **98**(10), 100133. Available from: https://linkinghub.elsevier.com/retrieve/pii/S0019452221001333 (accessed 22 November 2021).

Muzyka, R., Chrubasik, M., Stelmach, S. & Sajdak, M. 2015 Preliminary studies on the treatment of wastewater from biomass gasification. *Waste Management* **44**, 135–146. Available from: https://linkinghub.elsevier.com/retrieve/pii/S0956053X15300209 (accessed 11 August 2021).

Nidheesh, P. V., Couras, C., Karim, A. V. & Nadais, H. 2021 A review of integrated advanced oxidation processes and biological processes for organic pollutant removal. *Chemical Engineering Communications*, 1–43. Available from: https://www.tandfonline.com/doi/full/10.1080/00986445.2020.1864626 (accessed 27 January 2022).

OECD/OCDE 2010 Test no. 209: activated sludge, respiration inhibition test (carbon and ammonium oxidation). OECD Guidel. Test. Chem. Sect. 2 Eff. Biot. Syst https://doi.org/10.1787/9789264070080-en.

Philips, S. 2003 Impact of iron salts on activated sludge and interaction with nitrite or nitrate. *Bioresource Technology* **88** (3), 229–239. Available from: https://linkinghub.elsevier.com/retrieve/pii/S0960852402003140 (accessed 27 September 2021).

Pignatello, J. J., Oliveros, E. & MacKay, A. 2006 Advanced oxidation processes for organic contaminant destruction based on the Fenton reaction and related chemistry. *Critical Reviews in Environmental Science and Technology* **36** (1), 1–84. Available from: http://www. tandfonline.com/doi/abs/10.1080/10643380500326564 (accessed 10 July 2020).

Ribeiro, J. P. & Nunes, M. I. 2021 Recent trends and developments in Fenton processes for industrial wastewater treatment – a critical review. *Environmental Research* **197**, 110957. Available from: https://linkinghub.elsevier.com/retrieve/pii/S0013935121002516 (accessed 27 September 2021).

Sansaniwal, S. K., Pal, K., Rosen, M. A. & Tyagi, S. K. 2017 Recent advances in the development of biomass gasification technology: a comprehensive review. *Renewable and Sustainable Energy Reviews* **72**, 363–384. Available from: https://linkinghub.elsevier.com/ retrieve/pii/S1364032117300394 (accessed 19 November 2021).

Sharma, A., Ahmad, J. & Flora, S. J. S. 2018 Application of advanced oxidation processes and toxicity assessment of transformation products. *Environmental Research* **167**, 223–233. Available from: https://linkinghub.elsevier.com/retrieve/pii/S0013935118303748 (accessed 13 August 2021).

Su, X., Li, X., Ma, L. & Fan, J. 2019 Formation and transformation of schwertmannite in the classic Fenton process. *Journal of Environmental Sciences* **82**, 145–154. Available from: https://linkinghub.elsevier.com/retrieve/pii/S1001074218336702 (accessed 29 January 2022).

Tong, S., Shen, J., Jiang, X., Li, J., Sun, X., Xu, Z. & Chen, D. 2021 Recycle of Fenton sludge through one-step synthesis of aminated magnetic hydrochar for Pb^{2+} removal from wastewater. *Journal of Hazardous Materials* **406**, 124581. Available from: https://linkinghub.elsevier. com/retrieve/pii/S0304389420325711 (accessed 31 January 2022).

Tripathi, L., Dubey, A. K., Gangil, S. & Singh, P. L. 2013 Waste water treatment of biomass based power plant. *International Conference on Global Scenario in Environment and Energy* **5** (2), 761–764.

Umor, N. A., Ismail, S., Abdullah, S., Huzaifah, M. H. R., Huzir, N. M., Mahmood, N. A. N. & Zahrim, A. Y. 2021 Zero waste management of spent mushroom compost. *Journal of Material Cycles and Waste Management* **23** (5), 1726–1736. Available from: https://link.springer. com/10.1007/s10163-021-01250-3 (accessed 10 August 2021).

Usman, M., Ren, S., Ji, M., O-Thong, S., Qian, Y., Luo, G. & Zhang, S. 2020 Characterization and biogas production potentials of aqueous phase produced from hydrothermal carbonization of biomass – major components and their binary mixtures. *Chemical Engineering Journal* **388**, 124201. Available from: https://linkinghub.elsevier.com/retrieve/pii/S1385894720301923 (accessed 27 January 2022).

Wang, M., Zhao, Z. & Zhang, Y. 2019 Disposal of Fenton sludge with anaerobic digestion and the roles of humic acids involved in Fenton sludge. *Water Research* **163**, 114900. Available from: https://linkinghub.elsevier.com/retrieve/pii/S0043135419306748 (accessed 31 January 2022).

Wang, Y., Wang, H., Jin, H., Zhou, X. & Chen, H. 2022 Application of Fenton sludge coupled hydrolysis acidification in pretreatment of wastewater containing PVA: performance and mechanisms. *Journal of Environmental Management* **304**, 114305. Available from: https://www.sciencedirect.com/science/article/pii/S0301479721023677 (accessed 31 January 2022).

Widjaya, E. R., Chen, G., Bowtell, L. & Hills, C. 2018 Gasification of non-woody biomass: a literature review. *Renewable and Sustainable Energy Reviews* **89**, 184–193. Available from: https://linkinghub.elsevier.com/retrieve/pii/S1364032118301138 (accessed 19 November 2021).

Williams, B. C., McMullan, J. T. & McCahey, S. 2001 An initial assessment of spent mushroom compost as a potential energy feedstock. *Bioresource Technology* **79**, 227–230.

Wu, C., Chen, W., Gu, Z. & Li, Q. 2021 A review of the characteristics of Fenton and ozonation systems in landfill leachate treatment. *Science of The Total Environment* **762**, 143131. Available from: https://linkinghub.elsevier.com/retrieve/pii/S0048969720366614 (accessed 31 January 2022).

Zhang, Y. & Zhou, M. 2019 A critical review of the application of chelating agents to enable Fenton and Fenton-like reactions at high pH values. *Journal of Hazardous Materials* **362**, 436–450. https://linkinghub.elsevier.com/retrieve/pii/S0304389418308306 (accessed 27 January 2022).

Zhang, J., Liu, J., Evrendilek, F., Zhang, X. & Buyukada, M. 2019a TG-FTIR and Py-GC/MS analyses of pyrolysis behaviors and products of cattle manure in CO_2 and N_2 atmospheres: kinetic, thermodynamic, and machine-learning models. *Energy Conversion and Management* **195**, 346–359. Available from: https://linkinghub.elsevier.com/retrieve/pii/S0196890419305679 (accessed 11 August 2021).

Zhang, M., Dong, H., Zhao, L., Wang, D. & Meng, D. 2019b A review on Fenton process for organic wastewater treatment based on optimization perspective. *Science of The Total Environment* **670**, 110–121. Available from: https://linkinghub.elsevier.com/retrieve/pii/ S0048969719311684 (accessed 3 February 2022).

Zhao, Q., Han, H., Fang, F., Zhuang, H., Wang, D. & Li, K. 2015 Strategies for recovering inhibition caused by phenolic compounds in a short-cut nitrogen removal reactor treating coal gasification wastewater. *Journal of Water Reuse and Desalination* **5** (4), 569–578. Available from: https://iwaponline.com/jwrd/article/5/4/569/30276/Strategies-for-recovering-inhibition-caused-by (accessed 22 November 2021).

Zhu, H., Han, Y., Xu, C., Han, H. & Ma, W. 2018 Overview of the state of the art of processes and technical bottlenecks for coal gasification wastewater treatment. *Science of The Total Environment* **637–638**, 1108–1126. Available from: https://linkinghub.elsevier.com/retrieve/ pii/S0048969718316899 (accessed 12 August 2021).

First received 6 December 2021; accepted in revised form 12 February 2022. Available online 25 February 2022

IWA Publishing's authorised EU representative for General Product
Safety Regulations is Diane D'Arras, 15 rue Duret, 75116 Paris,
France, e-mail: safety@iwap.co.uk.

Printed and bound by CPI Group (UK) Ltd, Croydon, CR0 4YY
06/04/2026
02084337-0001